Fusion

A Voyage Through the Plasma Universe

Related Titles

Electron Acceleration in the Aurora and Beyond
D Bryant

Introduction to Plasma Physics
R J Goldston and P H Rutherford

Industrial Plasma Engineering
J Reece Roth

Plasma Physics Series

The Plasma Physics Series is an international series that meets the need for up-to-date texts in this rapidly developing field. Books in the series range in level from introductory monographs and practical handbooks to more advanced expositions of current research. Spanning both laboratory and astrophysical plasmas, topics covered include: magnetohydrodynamics and kinetic theory, waves and instabilities, magnetic and inertial confinement fusion, electron, ion and photon acceleration and heating, transport, turbulence, nonlinear plasma physics, dusty plasma physics, diagnostics, plasma processing and plasma simulation.

Series Editors

P E Stott, CEA Cadarache, France
H Wilhelmsson, Chalmers University of Technology, Sweden

New and Forthcoming Books in the Series

Plasma Physics via Computer Simulation, 2nd Edition
C K Birdsall and A B Langdon

Transport and Structural Formation in Plasmas
K Itoh, S-I Itoh and A Fukuyama

Instabilities in a Confined Plasma
A A Mikhailovskii

Nonlinear Instabilities in Plasmas and Hydrodynamics
S S Moiseev, V G Pungin and V N Oraevsky

Laser-Aided Diagnostics of Gases and Plasmas
K Muraoka and M Maeda

Inertial Confinement Fusion
S Pfalzner

Introduction to Dusty Plasma Physics
P K Shukla and N Rao

Collective Modes in Inhomogeneous Plasma: Kinetic and Advanced Fluid Theory
J Weiland

Fusion

A Voyage Through the Plasma Universe

Hans Wilhelmsson

Professor Emeritus
Chalmers University of Technology
Göteborg, Sweden

Institute of Physics Publishing
Bristol and Philadelphia

British Library Cataloguing-in-Publication Data
A catalogue record for this book is available from the British Library.

ISBN 0 7503 0639 4

Library of Congress Cataloging-in-Publication Data are available

Production Editor: Katie Pennicott
Production Control: Sarah Plenty and Jenny Troyano
Commissioning Editor: Michael Taylor
Editorial Assistant: Victoria Le Billon
Cover Design: Kevin Lowry
Marketing Executive: Colin Fenton

Published by Institute of Physics Publishing, wholly owned by The Institute of Physics, London

Institute of Physics Publishing, Dirac House, Temple Back, Bristol BS1 6BE, UK

US Office: Institute of Physics Publishing, Suite 1035, The Public Ledger Building, 150 South Independence Mall West, Philadelphia, PA 19106, USA

Typeset in TEX using the IOP Bookmaker Macros
Printed in Great Britain by Bookcraft (Bath) Ltd

Why fusion?

In the Universe
fusion became the source of energy
since it was the natural one
And
hydrogen became the fuel
since it was the simplest one
All stars and galaxies
were given this choice
Also the Sun
which provided conditions for
life on Earth
Why not choose fusion
to produce energy
for our society
And
build some new sources
by modern science and advanced technology?
They will be safe
And
have ecological advantages
And
they will be used forever
to the benefit of mankind
We are to prove that we can
realize this dream and that we can afford it

Contents

Preface

The term 'plasma universe' was coined by Hannes Alfvén to emphasize the decisive role of plasmas, or electrically conductive ionized gases, in our universe. Nuclear fusion is the fuel for all the stars and galaxies in space. Fusion therefore plays a very significant role. One could even talk about a 'fusion universe'. In the long term, fusion plasma may become the natural source of energy on Earth.

In this book we take a voyage through the universe and discover some remarkable and unexpected things. We will also examine some of the attempts to master the confinement of nuclear fusion plasmas in the laboratory. Many of the events observed in the laboratory can be tied in to discoveries that we make on our exciting voyage.

In this book I adopt a novel approach to fusion plasmas, covering fusion both in the laboratory and in the cosmos. I discuss the evolution of the field from early plasma research more than half a century ago to the gigantic scientific efforts of today covering basic fusion plasma science and laboratory fusion experiments, as well as geocosmophysical and astrophysical projects. The description is interspersed with passages which suggest relationships between science and art or poetry. I also relate meetings that I have had with famous scientists like Niels Bohr, Hannes Alfvén, Piotr Kapitza, Subrahmanyan Chandrasekhar, Lyman Spitzer and others.

There are several ways in which the similarities between fusion plasma phenomena in the laboratory and in outer space could be described. The natural connection comes from plasma physics, notably from the physics of waves and particles as well as of magnetic confinement of particles and energy in plasmas. Let us mention here magnetohydrodynamics (MHD), Alfvén waves, plasma waves, lower hybrid waves, shock waves and other nonlinear phenomena.

To understand the general outline of the book it should be emphasized from the very beginning that we must visualize two systems of fusion plant, each of a very different scale. One is the usual concept of a laboratory-designed thermonuclear reactor feeding an electric power station from which energy is distributed to consumers. The other type of fusion plant is the whole universe, which generates thermonuclear energy in the interior of the stars which distribute their energy, in the form of radiation (electromagnetic and particle radiation), to the rest of the cosmos and create the conditions for life on Earth.

In both systems plasma phenomena play a significant role. In fact, all phenomena of an electromagnetic and plasma nature which scientists try to use for confining and heating a plasma to thermonuclear conditions have counterparts in the natural universe, which is why I give many examples from astrophysics and geocosmophysics. Interestingly enough, very important progress in our understanding of numerous natural plasma phenomena is currently being made by the use of rockets and satellites. Simultaneously our knowledge of plasma phenomena in large artificial fusion devices is reaching a mature state.

A minor part of the text has been devoted to technology: it deserves to be included not only for its own sake but also to point out the relationships between science, art and poetry, and to bring us closer to reality.

Theory based on calculus has only been mentioned very briefly. Only a little notation describing physical quantities has been introduced. I indicate, however, that even trivial nonlinear relations can account for very significant phenomena governing nonlinear instabilities in plasmas or in other fields, for example population dynamics.

The book is essentially in two parts: the first part is devoted to plasmas and fusion in the cosmos, and the second to fusion on Earth. The first two chapters of Part 1 introduce the reader to the secrets of plasmas and to nuclear fusion reactions. The spectacular evolution from early history to present-day achievements in cosmic fusion and plasma physics is described in chapters 3 and 4 with the intention of giving a background for the topic of fusion on Earth in Part 2. An intermediate chapter 5 is devoted to electrical discharges and applications of plasmas, which are showing rapid development. Chapter 6 presents in a descriptive way the many efforts that have been made to form a rather consistent picture of the physical behaviour of fusion plasmas. Those readers who feel interested in penetrating the theory in more depth are referred to useful references. Chapters 7 and 8 treat recent developments in magnetic confinement and inertial confinement fusion, respectively. How a future reactor may be envisaged is touched upon in chapter 9. In chapter 10 the contents of the previous chapters are interrelated to give a basis for a look to the future and some conclusions. Finally a short fusion–plasma dictionary is given to elucidate the rich vocabulary used in my description of fusion plasma.

In writing this book I wanted to present to the reader certain points and ideas which have not been widely emphasized. The style is mainly popular, but the content is based on scientific analyses and observations. It is supported by a wide variety of illustrations in the form of photographs and drawings.

The book should be of interest to anyone with some curiosity about fusion and plasma physics and their current state of development, and should be accessible to readers with a fairly basic physics background from beginning undergraduate or even senior school level upwards. It emphasizes the very considerable achievements which have been made towards the use of fusion as the large-scale energy source for our society in the future.

Introduction

Stars, stars, stars ... wherever we look throughout space. They have been shining for ever, from the very beginning of the universe, and will continue to shine for billions of years. The source of energy for these stars is a process known as nuclear fusion. What an attractive source to benefit from if we could only learn to handle its enormous power.

The natural resources which we make use of today, such as coal, oil and gas, are all limited on Earth, as is the uranium needed for nuclear power reactors. The Earth's population is growing at a rate corresponding to an increase of two or three times its present size by the end of the 21st century. For anyone interested in the future evolution of our society an essential question is how to satisfy the need for a continuously growing energy consumption. It seems, indeed, that in the future fusion could provide our large-scale energy supply on Earth as it does already for the natural universe. It would also offer great ecological and safety advantages.

Fusion is directly related to what is called plasma, the hot ionized electrically conducting gas in which the fusion processes occur by reactions between light nuclei, e.g. hydrogen isotopes. The universe consists of more than 99% plasma, and hence there are obvious reasons why fusion and plasma science should be of the utmost general interest. The dream of an energy source based on nuclear fusion on Earth has stimulated significant activities in fusion research since the middle of the 20th century. Many nations today support large-scale fusion experiments. They all have the same goal—to harness nuclear fusion energy for the production of electricity. Important progress in fusion plasma science and technology has been made through the years but there is still a long way to go before a final version of a practical fusion reactor can be established.

Fusion reactions occur when nuclei of light atoms fuse together under the influence of nuclear forces. The hydrogen isotopes of deuterium (D) and tritium (T) are the most relevant ones to consider for fusion energy production. The state of matter in which the reactions occur is a hot plasma, where at high temperatures the atoms have split into free electrons and free nuclei which move as intermixed gases with opposite electric polarities.

For the fusion processes to become efficient, exceedingly high temperatures (10^8 kelvin) and sufficiently high densities (10^{14} cm^{-3}) are required. Also, the plasma has to be maintained for sufficiently long that energy has time to be produced before the plasma disappears due to destructive mechanisms (instabilities) or losses (radiation).

The dynamic behaviour of the plasma itself plays an essential role. The fusion reaction processes and the plasma dynamic processes (diffusion of particles and conduction of temperature) are mutually dependent—together they determine how the plasma will evolve in space and time.

The presence of plasma in all parts of the universe, in stars, galaxies, plasma jets and geocosmological environments, causes a close relationship between plasma physics and astrophysics, where recently extraordinary discoveries and technological developments (light and x-ray observations by space telescopes) have been made. The many spectacular plasma phenomena exhibited in large-scale universal contexts, as well as a detailed understanding of these phenomena, are of great interest for the interpretation of experimental results obtained in laboratory fusion plasmas.

Examples of spectacular plasma phenomena, e.g. plasma streams, comets, waves and instabilities, whirls and turbulences, are abundant in the universe. These phenomena may be caused directly or indirectly by fusion reactions. Similarities between those large-scale phenomena and events occurring at a much smaller scale in laboratory plasmas are of basic interest in plasma science. Also, comparisons with fluid phenomena in hydrodynamics, often studied by computer simulation experiments, give an insight into the behaviour of the more complex plasma systems.

Until recently, practically all fusion research has dealt with plasmas in the absence of fusion processes. In the presence of strong fusion activity (which requires advanced technological developments and which makes the interior of the fusion plasma device radioactive) the plasma behaviour may be considerably modified. The resulting heat and particle diffusion of the burning plasma affects the possibility of realizing a future reactor for electric energy production.

The new results and extensive experience obtained in laboratory experiments are important not only for the future planning of fusion reactors but also for understanding the results of observations in astroplasmas and geocosmophysical plasmas.

Ancient cultures used fire and flames to heat and prepare materials. Around 7000 years BC fire was used to make pottery and later on objects of copper, silver and gold. In the Middle Ages smiths made tools, swords and armour out of iron using fire and ovens to produce the heat. In early days flames were considered to be signs of purification and were even used to burn witches. In Greek mythology, Prometheus, a Titan who stole fire from Olympus to give to mankind, was chained to a rock as a punishment; here an eagle tore at his liver until Hercules freed him. The myth of Prometheus inspired Aeschylus to write a tragedy 'The Chained Prometheus', and the British poet Shelley to write the lyric drama 'Prometheus Unbound' (1820). In a recent painting, 'Cosmos' (1997) by Pierre-Marie Brisson, one may imagine Prometheus close to the fire (see plate 1).

In the space age we are used to seeing rockets take off carried by jet engines accompanied by plumes of luminous gas. Space telescopes are taking magnificent pictures of plasma jets extending through many thousands of light years and exhibiting structures like filaments and islands of plasma. The jets are thought to be created by energetic activity in the centre of a galaxy, possibly of matter from a supermassive black hole: this matter which once came, at the beginning of the universe, as dilute clouds of hydrogen to form the fuel of the slowly burning stars—the fusion plasma. These fascinating aspects have attracted the interest of numerous scientists all over the world and have also engaged great authors and artists over the centuries.

The importance of words and of formulae

Through the years research has become more international; scientists travel more and more and so do plasma physicists. They go to international conferences, to workshops, to numerous meetings, where large international experiments are planned. Often they go to laboratories in foreign countries to discuss scientific questions, to work on problems of common interest and to give seminars. In this way there has been an extremely active and very fruitful collaboration going on for many years. The common understanding of important questions has increased considerably. Sometimes it has been advocated that such scientific interchanges have even helped to open the borders between countries, and to improve the understanding between East and West.

In scientific activities there are quite a number of things that one has to learn to improve communication between scientists of different nationalities. Having had the opportunity of being a member of the scientific community when it was considerably smaller than today, when lack of international experience was more evident and mistakes were more frequent, I shall mention some of my own impressions and recount some of the lessons I have learnt through the years.

As a research student I was fortunate to spend a rather long period in Copenhagen at the Niels Bohr Institute, which was at that time considered to be the Mecca of theoretical physics. Many of the most famous physicists in the world came to the Institute and numerous seminars on the frontiers of physics were given. Niels Bohr, the father of atomic physics, sometimes attended the seminars himself, puffing his impressive pipe. The seminars were quite outstanding, often rather informal. A particular type of seminar was called the circus. It was intended to be on new ideas, or even on rumours of what was going on in international physics research. The common language was 'broken English', spoken without hesitation. It was astonishing and often amusing to see how a mixture of broken English and complicated formulae could support each other in the debate on models of the physical world. Even for the masters, however, the route to glory was hard. Niels Bohr told that he had rewritten the manuscript for his famous publication on atomic theory (Bohr N 1913 *On the constitution of atoms and molecules*, parts I–III *Phil. Mag.* **26** 1) at least 14 times and that he thought that every time it became worse!

Many years later I was invited to Kiev in the Ukraine to do research and give seminars on plasma physics. The first morning a young physicist came to my office and said 'I want to work with you'. I responded 'fine' and tried to explain what kind of research I was doing. However, it became clear very soon that we had no spoken language in common. I did not speak Ukrainian and he did not speak English, not even 'broken English', nor any other European language, for various reasons. But he was gifted and could handle formulae and understand physics. He continued to speak Ukrainian more or less to himself, but we found that we could communicate by another useful common language—mathematics in terms of formulae. His enthusiasm and my interest in the science and in Ukrainian culture led to a fruitful collaboration and resulted in several publications from my stay in Kiev.

On one of my other visits to Kiev several years later, I arrived well prepared for a talk, with a considerable number of view-graphs and many slides, most of them filled with formulae and graphs. But it turned out that the overhead projector had broken down and the slide projector was not sufficiently strong to compensate for the Ukrainian light that came in through the windows, which had no curtains. So I used words, just simple words in English, to explain for three hours, with an interpreter, what I had intended to communicate. It was on plasmas with nonlinearities, on wave interactions, on lasers and on physical models. And it became, under the circumstances, a great success, more so than it might have been with all the formulae on the slides!

This experience and some others stimulated me to write this book simply in words.

The vocabulary of fusion plasma science

The sea of words depicted below intends to convey the rich vocabulary used in fusion plasma formulation. All the words are given simple explanations in the short fusion–plasma dictionary at the end of the book.

UNIVERSE   CREATION   PLASMA

LIFE   MATTER

ALPHA PARTICLE   CHAOS   REACTION

FUTURE   STARS   COHERENCE

DEUTERIUM   SUN

NEUTRON   DENSITY   DIFFUSION

EXPLOSION   RADIATION   WAVE

LASER   SOURCE   SOLAR WIND

TURBULENCE   TEMPERATURE   ELECTRON

LIGHT   EQUILIBRIUM   ERUPTION

ORIGIN   EARTH   TOKAMAK

OSCILLATION   DIAGNOSTICS   IONS

STELLARATOR   POWER

INSTABILITY   ELECTRIC FIELD   STATE

NONLINEARITY   SUN SPOTS   PHOTON

AURORA   PROTON   GALAXIES

EVOLUTION   QUASAR   FLAME

SUPERNOVA   MAGNETIC FIELD   TIME

PULSAR   COMET   PROPAGATION

POPULATION   ENERGY

SKY   PROMINENCE   FUSION

SPACE   JET   TRITIUM   NEBULA

PART 1

FUSION IN THE COSMOS

Chapter 1

Plasmas

1.1 The secrets of plasmas

What mysteries lie behind all the beautiful objects that shine in the sky—those stars and galaxies, pulsars and nebulae, or the auroras, which produce glorious and colourful arcs in the northern regions of the globe?

The light comes from gases which occupy different regions of space and which exist under very different conditions. Ordinary gases like air at room temperature do not shine. They are not electrically conductive, which means that one cannot see any effect of a small electric voltage, like that from a battery for example, in the gas. However, if energy is transferred to the gas in order to heat it, for example by radiation, the atoms of the gas may be *ionized.* The electrons, which are bound by electric forces to the nuclei, will then be separated from their mother nuclei and they will fly away to other parts of the gas, to move around among all the nuclei. The gas then becomes a new state of matter which is electrically conductive and which is called a *plasma* [1.1–1.3].

If sufficient amounts of energy are transferred to the gas it becomes *fully ionized*, i.e. all the atoms of the gas will lose all their bound electrons. Before the atoms become ionized, the electrons could stay bound to their nuclei but in excited states. These have a higher energy than the original ground state. The electrons later decay from these excited states accompanied by the emission of radiation in the visible part of the spectrum. This happens, for example, in auroras where the atoms high up in the atmosphere become excited by electrons, which enter from the solar wind and become trapped by the Earth's magnetic field in the polar region. The decay of the excited atoms may subsequently generate beautiful luminous curtains of auroral radiation.

In astrophysical plasmas as well as in plasmas studied in the laboratory the motion of electrically charged particles, electrons and ions, are influenced by the presence of *magnetic fields*. The charged particles perform spiralling motions around the magnetic field lines. Magnetic fields could be generated by currents in the plasma itself, i.e. be self-generated, or could be generated by external magnetic field coils carrying currents generated in the laboratory [1.2].

In certain astrophysical objects like the Crab nebula a particular radiation, named synchrotron radiation, is generated by highly energetic, relativistic electrons which spiral around magnetic field lines generated by electron currents in the nebula itself. The beautiful visible structure of the nebula originates from this basic mechanism. Another phenomenon where the magnetic field has a confining effect on a plasma is that where plasma blobs are thrown out from the solar surface, carrying with them circulating currents which themselves create magnetic fields in the plasma by which the plasma stays confined. Plasma blobs could in this way be carried all the way to the Earth.

The state of a plasma has sometimes been called the fourth state of matter to emphasize its special nature.

Plasma occupies the following place among the various states of matter. In heating successively from low to high temperature one may identify various transitions between the different states of aggregation, namely:

(a) The solid state.

(b) When a solid is heated sufficiently it becomes a liquid. This occurs when the thermal motion of the atoms breaks the crystal lattice of the solid.

(c) When a liquid is heated to such an extent that its atoms become vaporized from a surface faster than condensation occurs on the surface it becomes a gas.

(d) At temperatures sufficiently high to ionize the gas, i.e. breaking the electron–nucleon bonds by collisions between the atoms in the gas so that the electrons become free, a plasma is created.

For plasmas heated to extremely high temperatures (hundreds of millions of degrees) nuclear fusion reactions may occur. The thermal velocities of the nuclei then become so high that the nuclei sometimes have a chance to approach each other so closely that the short-range attractive nuclear forces come into play and compensate for the electrically repelling forces. The nuclei then fuse together and enormous power becomes liberated.

Various combinations of hydrogen isotopes may undergo such reactions in stars and in fusion reactor plasmas. Self-sustaining burn conditions could then be achieved and the fusion plasmas could be efficient energy producing units for a long time.

Such burning plasmas are the origin of energy production in the stars, including our Sun (14–16 million degrees central temperature), which provides conditions for life on Earth. They will also be the source of energy production in future fusion reactors (200 million degrees burning temperature).

The classification of the elements given here is almost identical to that introduced by the Yin–Yang philosophy from ancient China, namely earth, water, air and fire. As a curiosity in this connection it may be mentioned that the noble coat of arms of Niels Bohr in Roskilde Cathedral in Denmark carries the Yin–Yang sign. In the coat of arms of Niels Bohr the Yin–Yang sign signifies the concept of complementarity in physics, i.e. the fact that an electron can behave like a particle under certain circumstances whereas in other connections it exhibits a wave-like nature. Complementarity is analysed on philosophical grounds in the literature of Søren Kierkegaard (1813–55) the great Danish author, whose ideas inspired Niels Bohr. The French physicist Louis de Broglie (1892–1987) also did fundamental work in pioneering the implications of the principle of wave–particle duality in atomic physics (and was awarded the Nobel prize in 1929).

As students in Sweden we were taught about Louis de Broglie and his work and we learnt that he lived in a beautiful French castle. I recently had the opportunity of visiting the castle, Chaumont-sur-Loire, belonging to the de Broglie family and happened to find that is was situated only some kilometres from another famous Loire Valley castle, Amboise, which belonged to the French Royal family and where Leonardo da Vinci spent the four last years of his life. There is now also an impressive museum in the city of Amboise exhibiting the many inventions by da Vinci. One cannot help thinking that new developments in the architecture of fusion plasma devices might benefit from the imagination of Leonardo (see chapter 7).

1.2   Plasma physics peculiarities and games

Plasma physics is a comparatively young science. Although some experiments relating to what are today called plasmas were done back in the 18th century, it was not until some decades into the 20th century that plasma physics started to develop as a separate science. The American scientist Irving Langmuir pioneered the field. He discovered that oscillations, so-called plasma oscillations, could occur in a plasma at a particular frequency, the plasma frequency v_p, determined by the density of free electrons in the plasma, n_e with v_p^2 being proportional to n_e. Langmuir also introduced the term 'plasma' and was awarded the Nobel prize for chemistry in 1932.

Plasma oscillation can be visualized as follows. The electrons in the plasma are repelled from each other due to their negative charges and adopt an equilibrium position. Suppose that we move a bunch of them by brute force in a certain direction and leave them suddenly. Restoring forces from the electric charges will then be set up and try to bring them back towards their original equilibrium positions. As a result they will swing back, but beyond their equilibrium positions due to their moments of inertia. They will subsequently return from the opposite side like a swinging pendulum and the process will be repeated.

Imagine another model from daily life: compare the electrons with cars which move on a road at a certain speed with a certain distance between them. For some reason one of the cars brakes. To avoid collisions the following cars will also brake, and so on, until the first car decides to recover its earlier speed followed by the others. The processes may be repeated. Along the line of traffic there will be a bunching of cars accompanied by a depletion of the density of cars. Motions of this character, longitudinal oscillations and waves, occur frequently in plasmas as in mechanical systems. The accompanying electric oscillating fields are obviously in the direction of the motion.

Electromagnetic waves, like microwaves, which are transverse waves with their electric and magnetic fields oscillating at right angles to the direction of propagation, become modified when they penetrate a plasma and change their velocity of propagation due to the presence of the plasma.

In the presence of static (non-oscillating) magnetic fields in the plasma a rich variety of waves of different character may occur. These may strongly affect the properties of the plasma. If the amplitudes of the waves increase to high values the waves may influence the transport of particles and temperature in a fusion plasma. Turbulent and chaotic phenomena in this connection are topics of high interest in today's plasma physics research.

There are certain low-frequency waves, for which the dynamics of the ions in the presence of a magnetic field in the plasma are important. They play a significant role in astrophysics as well as in laboratory plasmas. The pioneering work of the Swedish scientist Hannes Alfvén should be emphasized here. He was the creator of the field named magnetohydrodynamics and he greatly stimulated plasma-astrophysics in general. Alfvén was awarded the Nobel prize for physics in 1970. So-called Alfvén waves are low-frequency waves in magnetized plasmas. They can simply be simulated by the motion of an oscillatory suspended string. The analogy can be used to calculate the velocity of an Alfvén wave [1.4].

Hannes Alfvén's pioneering one page publication 'Existence of electromagnetic-hydrodynamic waves' published in *Nature* in 1942 is reproduced in appendix 1. It was followed by the first simple experimental demonstration of magnetohydrodynamic waves, also in one page, by Stig Lundquist (*Nature* in 1949) which is reproduced in appendix 2.

The physics of plasmas is nowadays studied by experimental, analytical and computer simulation methods. Experience has shown that plasmas are very complex media. The interplay between the different methods of investigation has, however, revealed important results even for the most complicated plasma structures. A presentation in terms of analytic formulae is, however, so lengthy and cumbersome that it is far beyond the aim of this discussion. Instead, it seems that comparisons with certain other activities and with games may be used to model, or at least crudely visualize, some features of plasma behaviour in complex situations.

The motions of the particles, or groups of particles, in plasmas may be compared with:

(a) Ordered motions like ballet or square-dancing. The motions of the particles in a plasma may be compared with the dancers of a modern ballet, where numerous dancers participate. The ballet 'Notre Dame de Paris' by the great French choreographer Roland Petit (La Scala, Milan, April 1998) was a marvellous performance of individual and collective motions, order and disorder, oscillation and waves stimulated by the music. Flames and sparks seemed to come from the vibration of the fingers of the dancers in coherence with the music.

(b) Collective drag, like a rugby team in action.

(c) Ice-hockey, for high-speed behaviour.

(d) Individual and collective motions of a set of chess pieces (slow motion).

(e) A tennis player in action, whose traces on the court would correspond to projections of the particle trajectories in the plasma. The ball could simulate an intermediary force. The lines on the court and the rules of the game might simulate the fusion device geometry and the influence of the confining magnetic field. There could be singles or doubles competitions depending on the configuration of the plasma, etc. The umpire of the competition would be the scientist, who ensures that the laws of Nature apply and that they are skilfully used to win the game!

In parenthesis it might be mentioned that games and sports seem to have been much in the minds of famous scientists: Niels Bohr, the father of atomic theory, and his brother Harald, who pioneered the field of almost periodic functions, were both prominent football players. Harald was in the Danish national team. Lyman Spitzer, pioneer of the American fusion programme was also an active mountaineer. Rumour has it that he got some of his ideas about particles in fusion plasma configurations (figure-of-eight configurations) from traces he made in the snow when skiing on the Matterhorn, the mountain which gave its name to the early American fusion program, Project Matterhorn.

References

[1.1] Spitzer L 1962 *Physics of Fully Ionized Gases* (New York: Interscience)

[1.2] Sturrock P A 1994 *Plasma Physics* (Cambridge: Cambridge University Press)

[1.3] Goldston R J and Rutherford P H 1995 *Introduction to Plasma Physics* (Bristol: Institute of Physics Publishing)

[1.4] Cross R 1988 *An Introduction to Alfvén Waves* (Bristol: Institute of Physics Publishing)

Chapter 2

Nuclear fusion

2.1 Nuclear fusion reactions

Nuclear energy can be liberated by two different processes, fission and fusion. In *fission* a heavy nucleus (uranium, plutonium, etc) is split into two or more fragments by bombardment of neutrons, simultaneously producing large amounts of energy. If uranium-235, consisting of 143 neutrons and 92 protons, is bombarded it produces 200 MeV for each nucleus that undergoes fission, two or three neutrons and also two radioactive nuclei. *Fusion* occurs when two light nuclei of hydrogen isotopes unite and undergo a fusion reaction from which neutrons of several MeV are emitted. For reaction of the hydrogen isotopes deuterium and tritium, the most studied process, a 14.1 MeV neutron is produced, and in addition an alpha particle, an He-4 nucleus, of 3.5 MeV. In the process an intermediate compound nucleus of two protons and three neutrons is formed which is unstable and immediately splits into a neutron and an alpha particle.

When do nuclear fusion reactions occur in a plasma? They can only occur when the temperature is very high, many millions of degrees. The reason is that the repulsion which always exists between the positive electric charges of colliding nuclei has to be overcome by attractive nuclear forces. This can only happen when nuclei with high mutual velocity come within the grasp of the strong but short-range (10^{-13} cm) nuclear forces, which occurs only for enormously high plasma temperatures—about 200 million degrees for deuterium–tritium reactions.

Nuclear fusion reactions can also be produced in a gas containing only deuterium; this would, however, require even higher temperatures, about 10 times higher than for deuterium–tritium reactions, considering burning conditions for a reactor.

Deuterium can be produced from ordinary sea water. The use of deuterium alone as fuel in a reactor would accordingly allow us to tap into an inexhaustible source of energy. Tritium, on the other hand, is a radioactive gas (with a half-life of 12 years) which can be generated from lithium by reactions with neutrons, directly in a reactor. Natural resources of lithium on Earth are estimated to be sufficient for tens of thousands of years.

A large-scale mechanical analogy of nuclear fusion would be to consider a gigantic rolling stone that is given a sufficiently high up-hill velocity to mount a volcano and fall into the crater to fuse with the magma!

To achieve nuclear fusion it is clear that the natural choice of nuclei would be to try the simplest ones, namely hydrogen or isotopes of hydrogen, which have the lowest electric charge to compensate and therefore require the lowest temperature. For fusion plasmas the nuclei and light particles (electrons, etc) participating in the fusion reactions are shown in table 2.1.

Table 2.1. Particles participating in fusion reactions.

proton, p ⊕	electron, e^-
neutron, n ○	positron, e^+
deuterium, D ○⊕	neutrino, ν
tritium, T ○	gamma, γ
○⊕	
helium 4, ^{4}He (alpha particle) ⊕○	
○⊕	
helium 3, ^{3}He ⊕	
○⊕	

It might be interesting to see how Nature has solved the fusion problem in the stars, notably in the Sun, which is responsible for life on Earth and which is the centre of our own solar system. It has been determined by H Bethe (who won the Nobel prize for physics in 1967), among others, that the nuclear fusion reactions which provide the energy in the Sun, and essentially in all stars, belong to a cycle of proton–proton reactions. The first step in the sequence is the one where the reaction between two protons produces one deuterium nucleus and in addition one positron, one neutrino and gamma radiation. This type of reaction liberates 9 MeV of energy. The deuterium produced reacts with a proton to form a helium-3 nucleus and in addition gamma radiation. The liberated energy amounts to 5.5 MeV. Finally two helium-3 nuclei react to reproduce two protons like the ones that started the chain, plus a helium-4 and further gamma radiation, the process liberating 2.8 MeV. The reactions require a temperature of 6 million degrees, whereas according to available theory the centre of the Sun has a temperature of 14 million degrees. This chain of nuclear processes is summarized below using the following notation: p denotes a proton, D is deuterium, e^+ a positron, ν is a neutrino, γ is gamma radiation and ^{3}He and ^{4}He are helium-3 and helium-4 nuclei respectively.

Step 1: $p + p \rightarrow D + e^+ + \nu + \gamma$ (9 MeV liberated)
Step 2: $D + p \rightarrow {}^3He + \gamma$ (5.5 MeV liberated)
Step 3: $^3He + {}^3He \rightarrow {}^4He + p + p + \gamma$ (2.8 MeV liberated).

The energy carried by the gamma rays is finally transformed into visible light etc after a long complex journey through the Sun.

2.2   Nuclear fusion plasmas

What else is there that is important and characteristic in fusion plasmas? There is a basic phenomenon which has to do with the property of all plasmas to attempt to be electrically neutral on a scale larger than a certain value, called the Debye distance (named after Peter Debye who was awarded the Nobel prize for chemistry in 1936).

The square of the Debye distance, denoted λ_D, is proportional to the ratio between the plasma temperature (T) and the density of free electrons (n), i.e. $\lambda_D^2 \sim T/n$ [2.1]. If one considers free electrons, which are attracted to some nuclei, there is compensation for the influence of this attraction by the thermal motion of the electrons. As a result shielding of the positive charges by the electrons occurs, such that outside a very small distance from the positive charges, the plasma is electrically neutral. This means that the positive charges are electrically hidden to an observer outside the Debye distance, which is of the order of or less than a few millimetres in a fusion plasma. The concept of a neutral plasma has a meaning only for plasmas larger than the Debye distance. For comparison, it could be mentioned that rocket flame plasmas could have cross sections of the order of the Debye distance and might therefore not be considered as neutral plasmas.

A fusion plasma, being a fully ionized gas, has certain similarities with the properties of a metal. The abundance of free electrons is the reason why the plasma is a good electric and thermal conductor. The total energy W liberated by fusion per unit volume of a plasma of temperature T is proportional to the product of the density of the two reacting nuclei n_1, and n_2 and to the energy Q liberated in each reaction as well as to a factor P, the probability of reaction, which depends on T and accounts for how often a nucleus of one type has a fusion reaction with a nucleus of the other type.

In the energy budget of the plasma one has to take into account losses by various mechanisms such as electromagnetic radiation and thermal conduction. As a result heat energy will diffuse away from regions of high temperature, where fusion reactions occur most frequently. Fine structure effects like turbulence and instabilities may considerably change the diffusion from what would be expected from elementary classic theory [2.2]. The diffusion of particles and the conduction of heat may be considerably reduced by the influence of a confining magnetic field generated by currents external to the plasma or by currents in the plasma itself, as in the tokamak experiments, the most successful ones to date.

It seems that in present fusion plasma research we may be able to master the confinement problem by using magnetic fields. The process of magnetic confinement originates from the phenomenon that electrically charged particles tend to perform spiral motions around the magnetic field lines which define the direction of the magnetic field in each point in space. Thus, the particles will not be free to penetrate space perpendicular to the direction of the magnetic fields.

Dangerous effects similar to the explosion of an atomic bomb, which uses the same fusion reactions, cannot occur in a fusion reactor plasma. The metallic walls of the vessel surrounding the plasma would cool the expanding plasma in a short while.

Another interesting alternative for producing fusion plasmas, in addition to magnetic confinement, has attracted considerable attention for a long time, namely the process of inertial confinement. In essence the principle amounts to performing the experiment so quickly that the plasma has virtually no time to expand because of inertial forces. The energy for heating the solid materials used in the experiment, in fact small ice pellets, to fusion temperatures is supplied by pulses of high power laser radiation or of ion beam pulses with appropriate repetition rates.

It is still an open question as to which of the two principal confinement schemes, magnetic confinement or inertial confinement, is the more attractive or more realistic. The inertial confinement scheme is highly dependent on the evolution of high power lasers, such as those which are currently being developed in the megajoule energy domain [2.3, 2.4].

The successful operation of fusion reactors using either of the two main confinement schemes relies strongly on the simultaneous optimization of the complex nuclear fusion plasma and the surrounding active auxiliary devices (plasma geometry and size, heating systems, superconductive coils, etc). But how is the work carried out? What do the plasma physicists do? There is a story that goes back to professor Victor Weisskopf, when he was Director General of the European Organization for Nuclear Research (CERN). He was asked one day about what physicists in general do. We may here transform his answer to explain what a plasma physicist does: there are different kinds of plasma physicists. Some work close to engineers, building big devices like accelerators or fusion machines. Let us compare them with shipbuilders constructing big boats. Then there are those who use the devices to make new experiments and discover important things. They lecture to their students and report about their work as they go around the world to meet other plasma physicists and discuss their results. They are to be compared with sailors who set their sails to go across the oceans to meet many difficulties and perhaps discover new continents. But then there are also plasma theoreticians. They are to be compared with the people who remained in Spain to make calculations to prepare and guide Christopher Columbus for his voyage to India but who made him end up in America! (see also chapter 6).

References

[2.1] Chen F F 1984 *Introduction to Plasma Physics* (Dordrecht: Kluwer)
[2.2] Miyamoto K 1976 *Plasma Physics for Nuclear Fusion* (Cambridge, MA: MIT Press)
[2.3] Herman R 1991 *Fusion: The Search for Endless Energy* (Cambridge: Cambridge University Press)
[2.4] Fowler T K 1997 *The Fusion Quest* (Baltimore, MD: Johns Hopkins University Press)

Chapter 3

The cosmos

3.1 Fusion in the cosmos

Messages about new scientific discoveries or new technological achievements reach us frequently. Often these come from the field of astronomy or space science. New instruments of observation have been developed using rockets, satellites or space shuttles. The exploration of our solar system has opened up a new era of scientific investigation. The development of data communication techniques has made it possible for us on Earth to stay in contact not only with astronauts making more and more distant trajectories in their spacecraft, but also with observing instruments and recorders on unmanned spacecraft sailing further and further out in the archipelago of remote planets, passing through belts of asteroids and the continuously blowing solar wind, measuring parameters of distant magnetospheres and studying magnetic fields and space plasmas.

Observations of galactic systems and other extremely remote sources like quasars offer great challenges. Large optical telescopes and radioastronomy antennas are detecting radiation and signals from the outer limits of the universe. In this enormous stream of fascinating news and observations on cosmic events we all remain full of admiration for scientific progress. But seldom are questions raised about what is driving this gigantic machine of celestial objects—our universe. Where does the energy come from?

The unique answer is that the origin of the power driving the universe is fusion energy generated in a plasma. But how does this system evolve and how does the originally generated energy become distributed among the tremendous variety of stars and galaxies and other celestial objects that we find in our universe? Visual observations only give information about the light from the objects or gases which are kept together by gravitational forces. But what is the function behind the emission? What role do the magnetic fields and currents play in the process of evolution of the system? We are now in a position to find out much more about those questions. One may wonder why we have not already paid more attention to energy problems on a universal scale. One reason is that we cannot see or directly observe celestial fusion energy processes or plasma-magnetic phenomena occurring in space plasmas and remote magnetic fields. The frequencies associated with plasma-magnetic phenomena are simply not in the visible range but in the regions of centimetres, metres or even longer wavelengths. However, the possibility of clarifying the role of fusion and of plasma-magnetic field systems in the universe has increased tremendously in recent years. The situation provides a great challenge for surveying fusion plasma problems in the universe and for encouraging continued studies of problems common to both cosmic fusion plasma electromagnetics and device-oriented fusion plasma research aimed at production of electric power by fusion reactors on Earth.

3.2   How it all began . . .

The universe is considered to be about 15 billion years old. 'News' can be brought to us on Earth by light or radiowaves over distances of billions of light years. What we can see today, looking out into the universe with the naked eye or sophisticated instruments, is something exceedingly spectacular. It tells us that, since the beginning of the universe, Nature has played an extremely active game and created many structures. Each of them is evidence of a certain moment in the history of the universe. Dramatic events exhibiting many features of what we today call modern physics have occurred in gigantic dimensions at enormous distances. The story of the evolution of the universe is, in fact, recorded in the scenario around us, telling us about galaxies, stars, nebulae, supernovas and pulsars. One might even say that in a certain sense the scenario is a visualization of evolution in four-dimensional space. What a fantastic source of information!

What an extraordinary exhibition of organization and indeed self-organization of matter! How did it all come about?

The field is as open to speculation as it is fascinating and complex. To what extent can we hope to describe the very beginning of the universe?

We may even ask: was there really a beginning? Perhaps the universe has been always oscillating, creating and annihilating itself over a huge timescale. Perhaps it has a different evolution from what we can even imagine today.

We now think, or at least many astrophysicists think, that there *was* a beginning and that it was manifested by an event resembling some kind of enormous lightning flash which occurred simultaneously everywhere and that it resulted in a state of extremely high density and temperature. As the poet *may* have expressed it

This is the way
the world began
not with a whimper
but with a BANG

(with apologies to *T S Eliot, 1885–1965*).

We now think that during the very first moments of the universe there were no galaxies or stars, no atoms or nucleons, no structures whatsoever. Photons, electrons and quarks, as the building blocks of Nature, were somehow hidden in a uniform background, where there were also gluons ready to help the quarks (one talks about quark–gluon plasma) to combine in triples to form protons or neutrons. Each of these would then include two quarks of one type (charge 2/3) and one quark of another type (charge

−1/3), which would give a proton (charge 1), or include one quark of type one (charge 2/3) and two quarks of the other (charge −1/3) yielding a neutron (charge 0) [3.1, 3.2]

Returning to the initial instant of the universe the opinion is that during the very first minutes essentially only hydrogen and, as a result of fusion reactions, helium as well as isotopes of these elements existed. All other elements were formed much later, in fact some hundreds of millions of years later, in the interior of heavy stars. What happened to the light, the tremendous flash that accompanied the initial bang?

To answer this question it is necessary to emphasize another important point which has to do with the overall motion in the universe, which is believed to be characterized by expansion in every point of the universe. It means that wherever we imagine that we are in the universe we see all other points disappearing away from us, the near ones as well as the distant ones. Since this expansion is uniform everywhere in the universe it follows that the distant points move away from us with higher velocities, which are, in fact, proportional to the distance. In the same way as an implosion raises the temperature in a gas the expansion has the tendency of cooling not only the expanding matter but also the light, making the light less energetic. Since red light is less energetic than blue light it follows that one would expect the cooling effect from the expansion to shift the blue light more and more towards the red domain of the spectrum and to the invisible part of even longer wavelengths occupied by radio waves. The light which existed in the very beginning exists today in the form of radio waves in the whole universe as a background radiation, or as a 3 K radiation, referring to the temperature in kelvin of its black-body radiation. It can be observed by radio telescopes on Earth. The discovery in 1965 of the 3 K background radiation (−270 °C) by the radio-astronomers Arno Penzias and Robert Wilson led to the award of the Nobel prize for physics in 1978.

3 K radiation is remarkable in that it witnessed the very beginning of the universe. In his poem *Doriderna* Harry Martinson,

the great Swedish author (1904–78), describes the formation of the universe in the following way:

> *The soul of ideas in space*
> *out of infinite haze*
> *gathered the seeds*
> *to the Sun's patient blaze*
>
> *From far beyond time*
> *came the hydrogen*
> *out of its modest dress*
> *and built the atoms'*
> *ingenious nests to God bless*

(freely translated from the original Swedish!).

On your voyage through the universe you may, like Nils Holgersson (created by the Swedish novelist Selma Lagerlöf 1858–1940) on his marvellous voyage, travel on a goose-shaped space shuttle and land in areas where you find new and unexpected friends, revealing to you secrets about new landscapes of the universe; you may come across many regions which you find fascinating and perhaps others which you find less attractive. Like Gulliver (in Jonathan Swift's *Gulliver's Travels*), you may encounter domains and perhaps inhabitants which seem quite strange and of very different size, both giants and dwarfs. You will hear cosmic music and different signals whistling when you pass plasmas surrounding cosmic bodies. You will see gigantic plasma jets crossing through space and enormous eruptions of plasma emerging from the stars. The source behind all this is fusion energy—the immense cosmic burn-out of nuclear matter.

3.3   Galaxies and stars

In the universe galaxies are not distributed uniformly. They tend to form groups or clusters. Our galaxy is a member of such a group of about 30 galaxies. It is one of the largest in the group, to which also belongs the spiral galaxy Andromeda, the largest member of the group. Other types are elliptic, irregular and spiral-bar galaxies. Typical values of the diameters of the spiral galaxies are 100 thousand light years and the distances between them are about 10 million light years. As a comparison light takes eight minutes to go from the Sun to the Earth. An average galaxy contains many hundreds of billions of stars. The spiral arms of the galaxies are regions from where the stars originate. At a distance a spiral galaxy looks like a rotating firework. In reality the time of rotation of a galaxy amounts to some million years.

But where did the structures in the universe come from in the very beginning? One thinks today that out of the enormous temperatures and densities some kind of state emerged where certain regions contained slightly more matter than others and that gravitational forces which rule the universe led to concentrations of matter which subsequently developed to individual galaxies and stars. This process took a very long time. Even millions of years from the very first moment nothing but hydrogen and helium is thought to have existed in the universe in the form of gas clouds. The temperatures were still thousands of degrees. When structures started to develop the regions between the stars still contained matter, essentially hydrogen and helium, but also beginning forms of molecules and fine dust which came to form interstellar clouds and nebulae, visible in telescopes as gigantic formations, the environment of the origin of life [3.3–3.6].

In the history of cosmology it should be noted that it was in 1923 that the American astronomer Edwin Hubble showed that Andromeda was a separate galaxy far out of our Milky Way. In 1929 he showed that the galaxies moved away from each other. His name was given to the Hubble constant, the characteristic constant that describes the mutual velocity dependence on the distance of separation, and subsequently, to the extraordinary Hubble Space Telescope that has led to a revolution in optical astronomy and cosmic sciences.

In art and literature the wonders of the heavens have been great sources of challenge and inspiration. In some paintings by the great Dutch master Vincent Van Gogh (1863–90) the artist stresses the effects of light by emphasizing the contrast between the blue-green and the yellow colours. As a result he obtains effects like enhanced vibrations in the light (see plates 2 and 3). The cypresses in these paintings, given dark green and bronze colours, are also important contrasts to the galaxies and stars. Those cypresses take the form of fire-flames similar to those observed in eruptions from the solar surface. The contrasts as well as the strong and even mysterious impressions which the paintings give are furthermore accentuated by the typical stream-like patterns which Van Gogh produces in his paintings by the strokes of his paintbrush.

Ilia Zdanevitch (1894–1975), (after 1920 Iliazd), a poet from Tbilisi (Tiflis), Georgia, published in 1964 an extraordinary book illustrated by Max Ernst (1891–1976) *Maximiliana ou l'Exercice Illegal de l'Astronomie.* Iliazd's typography introduces the words like patterns of galaxies, comets and meteors exhibiting and relating the macro-cosmos and the micro-cosmos, even touching upon the creation of life.

Max Ernst invented for the book a special strange alphabet of hieroglyphs which give the pages the expression of invisible forces like those interlinking celestial bodies. The book recognizes the work of Guillaume Tempel, an autodidact in astronomy who in 1861 discovered a small planet between Mars and Jupiter. He gave it the name Maximiliana in honour of Maximilian II of Bavaria. He became, however, discredited by the official astronomers, who gave the name Cybèle to the planet and deprived Tempel of his discovery. Iliazd and Max Ernst, fascinated as they were by astronomy and cosmic events, decided to collaborate in the written language of the 'alphabet of the stars' and they succeeded in combining the text with splendid etchings to produce the magnificent book *Maximiliana* and to shed proper light on the achievements of Guillaume Tempel (see plate 4).

There are several other great authors, among them Victor Hugo (1802–85) ('La Comète', p 219 and 'Abîme', p 411 *La Légende des Siècles II* (Paris: Flammarion) 1967) who in their works have been inspired by cosmic sources.

Leonardo da Vinci (1452–1519) described and made drawings of how light was spread among the leaves of the trees and of reflections in the atmosphere 500 years ago, unbelievably in advance of his time. He also constructed eye-glasses and simple optical instruments as precursors of the astronomical tubes a hundred years before Galilei. When he writes: 'Where a flame cannot live no animal which breathes can live' he touches upon modern chemistry and the conditions of life [3.7].

Returning to physics, present-day research provides evidence of a new and unexpected feature, namely that there is an inherent anisotropy of the entire universe, i.e. that there is a preferred direction in the universe. The evidence comes from changes in the polarization of radiation, which has propagated over very long distances, and where the observed data cannot be explained by existing models of an isotropic universe. The indications have, however, to be examined in great detail and checked by more observations and statistical analyses of the results before safe conclusions can be drawn. If correct, the results could have important consequences for cosmology. There is apparently still room for exceedingly complex and far-reaching studies of the cosmos!

One may wonder what role interstellar magnetic fields and plasmas play in this connection. The Earth and the Sun, and presumably also all the other stars, have magnetic fields with preferred directions which change polarity periodically. The galaxies have individual spin directions of their vortex motions. So why could not our whole universe be anisotropic?

The first observations of anisotropies in the cosmic microwave radiation by the Cosmic Background Explorer Satellite (COBE) in 1992 had an important impact on research in the field of cosmology. The discovery of the anisotropies, followed by a large number of detailed studies using different types of observational technique, may give a clue to the formation of seeds in the universe that subsequently developed into galaxies and larger structures. It should be noticed that the relative anisotropies are tiny, one part in 100 000 as obtained by the COBE satellite. However, those small deviations could have extremely important consequences for the evolution of the universe.

3.4   The Sun's metabolism

The Sun is the only star where details, for example dimensions of less than some hundred kilometres, of the motions of ionized gas in the presence of magnetic fields can be studied by optical instruments. All other stars appear as point sources even in the largest telescopes, which can therefore only observe and analyse the properties of the emitted radiation. The mass of the Sun is 2×10^{33} g, or 330 000 times larger than the mass of the Earth. The solar diameter is 1.4×10^6 km which is more than 100 times larger than that of the Earth. The Sun is about 150 million km from the Earth, a distance which it takes eight minutes for light to travel. The surface temperature of the Sun is about 6000 °C.

For many thousands of millions of years the Sun has been emitting a total amount of radiation of 4×10^{23} kilowatts of energy per second in the form of light as electromagnetic waves, which corresponds to hundreds of billions times the power consumption of the whole of the United States. This would correspond to a kilowatt of power on every square metre of the Earth's surface exposed to the Sun's isotropic radiation. One knows that this radiation has been emitted for billions of years, the age of the oldest organic fossils found. The source of emitted energy lies deep in the Sun, where the temperature is about 15 million degrees. Optical observations can give us direct information about only a thin layer of the solar surface. To explain what happens inside the Sun one must rely on theoretical modelling. The models should account for everything that one knows about the Sun, in steady state as well as under perturbed conditions, i.e. in the presence of solar activity. As far as the steady-state situation is concerned one knows that there is a force of gravity at every point of the Sun which tends to contract the Sun, and also that there is a gas pressure which tends to expand the Sun. The force of gravity and the gas pressure have to balance each other to maintain equilibrium. The stable situation has remained for billions of years, and the size of the Sun and its brightness have been essentially unchanged.

In 1960 an oscillation of the Sun with a period of 5 minutes was discovered. Today it is known that the periods of oscillations of the Sun have values between 5 minutes and more than one hour. These oscillations can be used to analyse the internal structure of the Sun.

Only for central temperatures of tens of millions of degrees is the gas pressure sufficiently high to balance the enormous gravitational forces from the total volume of the Sun. The equilibrium pressure corresponds to a density of matter in the centre of the Sun of 100 grams per cubic centimetre, which is maintained in a region of one-quarter of the solar radius from the centre. This region contains almost half of the total solar mass, whereas one-quarter of the solar radius from the surface would define a shell where the density would be less than a tenth of a gram per cubic centimetre. It is in the central sphere where the energy is liberated by nuclear reactions transforming hydrogen into helium. The nuclear processes provide about one million kilowatt hours per five grams of helium, which would not occupy more than a few cubic millimetres.

This brings us to another point of fundamental interest, namely the equivalence between mass and energy as expressed by the Einstein equation $E = mc^2$, where E is the energy, m the mass and c the velocity of light (3×10^{10} cm s^{-1}). The equation tells us that one gram of mass is equivalent to 25×10^6 kilowatt hours. The liberation of energy by transmutation of protons and neutrons to helium is accompanied by a mass defect multiplied by the square of the velocity of light.

There is no other possible explanation for the enormous energy liberated by the Sun or the other stars than nuclear fusion. The form of hydrogen burning to helium in the initial phase of evolution is the natural solution to the problem of solar energy production. In fact, the Sun is essentially a gigantic fusion reactor. A fusion reactor on Earth is presumed to operate with tritium and deuterium, the isotopes of hydrogen, to be efficient at available temperatures (2×10^8 °C), whereas the Sun uses only hydrogen (protons) to generate the helium. At such high temperatures the atoms are always completely stripped of their electrons. When, during the passage through the Sun, the gamma rays, which are also produced by the nuclear reactions, carry the energy outwards from the inner central region and reach the outer quarter radius of the Sun, the energy will start to be carried further by convection. Like boiling water the solar matter starts to seethe and evaporate and then reaches the photosphere, a thin gaseous layer (10^{-16} grams per cubic centimetre) which has a granular structure of irregular convection cells.

Above the photosphere lies the chromosphere, which is more turbulent than the photosphere, and has a density of only 10^{-11} grams per cubic centimetre, which further out falls to 10^{-16} grams per cubic centimetre and suddenly drops even further and becomes the corona, a rarified layer of gas, which stretches out for millions of kilometres. The corona can be seen only during total eclipses when the Moon passes in front of the Sun to fully hide its luminous surface.

Returning to the problem of energy production in the centre of the Sun, the most important cycle of fusion reactions is

$$p + p \rightarrow D + e^+ + \nu + \gamma + 9\,\text{MeV}$$

$$D + p \rightarrow {}^3He + \gamma + 5.5\,\text{MeV}$$

$${}^3He + {}^3He \rightarrow {}^6Be \rightarrow {}^4He + p + p + \gamma + 2.8\,\text{MeV}$$

where the initial proton–proton reaction produces deuterium (and simultaneously a positron and a neutrino) and in the second step the deuterium reacts with a proton to produce a helium-3 isotope and a quantum of gamma radiation. The generated positron may annihilate with an electron to produce further gamma radiation ($e^+ + e \rightarrow 2\gamma$).

Finally, two helium isotopes react to produce helium and recover two protons after an intermediate step where beryllium

^{6}Be is formed. The net effect of the cycle is that one helium nucleus is produced by four hydrogen nuclei. Each step liberates several MeV. The gamma rays become transformed to visible light on their way through the Sun. The cycle of processes here described is what gives the Sun and the Earth their energies. For solar fusion to occur a temperature of 6 million degrees is required, whereas the centre of the Sun has a temperature of about 14 million degrees. For stars which are heavier than $2M_\odot$ ($M_\odot$ denotes the solar mass), for which the temperature in the centre becomes higher than 20 million degrees, another cycle where carbon, nitrogen and oxygen are involved will be responsible for the energy production [3.8, 3.9].

The reaction producing deuterium out of protons is fortunately very slow. The average reaction time per particle is estimated to be 1.2×10^{11} years in the centre of the Sun. This peaceful burning is the reason for the long lifetime of the Sun. The age of the Sun is about 5 billion years and it is estimated that the Sun has sufficient hydrogen fuel to burn for another 5 billion years. But what will happen then? The central part of the Sun will contract to heat and produce more energy whereas the outer parts will expand and cool. The colour of the radiation will change from white, via yellow to red. The Sun will become a red giant and later on a planetary nebula. Meanwhile the central part of helium will continue to contract, and when the increasing temperature reaches about 80 million degrees the helium will start to burn. The helium nuclei will undergo nuclear fusion reactions to produce carbon and radiation with an intermediary production of beryllium, ^{8}Be

$$^4\text{He} + {}^4\text{He} \rightarrow {}^8\text{Be}$$

$$^4\text{He} + {}^8\text{Be} \rightarrow {}^{12}\text{C} + \gamma.$$

The carbon may further react with helium to produce oxygen and gamma radiation

$$^4\text{He} + {}^{12}\text{C} \rightarrow {}^{16}\text{O} + \gamma.$$

The central part, which will now have a mass of about half the original total solar mass will form a so-called white dwarf star, which can be observed by telescope in the central region of planetary nebulae. The density of a white dwarf is about one ton per cubic centimetre.

Stars which are heavier than our Sun can continue to build up even heavier nuclei with masses which are multiples of the helium mass. The central part will, however, not have a mass higher than $1.4M_\odot$, the Chandrasekhar limit, determined by the fact that for the high pressures prevailing in the central star the electron gas does not behave normally but becomes degenerate. The Chandrasekhar limit is an interesting one conceptually, and has important astrophysical consequences. The result of detailed theoretical calculations shows that the Chandrasekhar mass is proportional to the factor $(hc/2\pi G)^{3/2}$, h being Planck's constant, c the velocity of light and G the gravitational constant. One notices the interesting interplay between the constants of quantum mechanics, relativity and gravitation in this dependence. It was obtained by Chandrasekhar while on a boat journey from India to England in 1930. Arriving there at 20 years of age he introduced his results to Sir Arthur Eddington, the leading British astronomer of the time and director of the Greenwich Observatory, who was then working on the radiation equilibrium of stars. The young Chandrasekhar succeeded in convincing Eddington about his result, and later, in 1983, he received the Nobel prize for physics.

References

[3.1] Hawking S W 1998 *A Brief History of Time* (New York: Bantam Books)

[3.2] Weinberg S 1977 *The First Three Minutes, a Modern View of the Origin of the Universe* (New York: Basic Books)

[3.3] Hoyle F 1975 *Astronomy and Cosmology, a Modern Course* (San Francisco: Freeman)

[3.4] Flammarion C 1880 *Astronomie Populaire* (Paris: Flammarion)

[3.5] Reeves H 1984 *Poussières d'Étoiles* (Paris: Editions du Seuil, Sciences)

[3.6] Vanin G 1995 *Astronomie Images de l'Univers* (Paris: Gründ)

[3.7] Bramly S 1988 *Leonard de Vinci* (Paris: Lattes)

[3.8] Kippenhahn R and Weigert A 1990 *Stellar Structure and Evolution* (New York: Springer)

[3.9] Clayton D D 1968 *Principles of Stellar Evolution and Nucleosynthesis* (New York: McGraw-Hill)

Chapter 4

The plasma universe

4.1 Around the Sun

4.1.1 The magnetoplasma revolution

Magnetic fields and electric currents play an important part in the workings of the cosmos. It has been thoroughly settled that many significant features in the make-up of the universe, like the physics of stars and galaxies or of the magnetospheres of stellar objects, are related or even governed by coupled electromagnetic–plasma fluid phenomena. Related questions are of vital interest in many different areas of research, for example in galactic and solar physics, as well as the physics of the magnetosphere, i.e. the physics of the magnetoplasma surrounding our planet.

In the beginning, the progress of cosmic electrodynamics was very slow. Only since the 1950s has it been seriously applied to astrophysics. Other classical fields of physics, like classical mechanics, spectroscopy and nuclear physics, had been applied extensively to astronomical problems for many decades. As a result much was already known about the motions of stars, their temperatures and chemical composition before electromagnetic fields and plasmas were even considered. Now we are in a position to take a journey around the universe, to see the impact of plasmas in their electromagnetic fields.

The fact that the Earth is a magnet has been known for some 400 years and the magnetic perturbations on Earth associated with auroras have been known for at least 200 years. Strong magnetic fields in Sun spots which could only be caused by gigantic electric currents on the Sun were discovered about 100 years ago.

Observations of radiowaves from the Sun as well as from other radio sources in the universe, in the late 1940s, marked the beginning of radioastronomy, which used wavelengths in the atmospheric 'window' of the frequency band, corresponding to wavelengths from a few millimetres to about 15 metres. The science of radioastronomy has since then grown tremendously in importance and in scope. It has led to many discoveries about physical conditions and chemical abundances of elements in the universe. Radioastronomy has contributed important information to cosmological theories. New sources of radiation in the cosmos like quasars (1960) and pulsars (1967) were discovered and studied using radio telescopes.

At the same time space technology has developed enormously, offering new platforms of observation outside the atmosphere like rockets and satellites carrying instruments for detailed exploration of the Earth's magnetosphere and its plasma surrounding. Our view of the plasma-magnetic configuration around our planet has accordingly been strongly modified and the role of the solar wind, i.e. the steady stream of protons and electrons from the Sun, has been elucidated in a very profound way. Bearing in mind that modern plasma laboratory experiments and fusion plasma physics have developed essentially within the same period of 50 years, since the 1950s, providing today an enormous integrated knowledge of plasma physics, the fields of plasma and fusion science have shown impressive growth.

As we have said, it took a long time before magnetic fields and electric currents became accepted as essential ingredients in the cosmos. This is still happening now. Why so late? Astronomy has been a field of great interest since antiquity, and particular events like solar eclipses, Sun spots, auroras etc have attracted considerable attention for centuries. I suggest that it was simply not natural to think of magnetic fields and electric currents as playing an important role in outer space. Electric currents are something that one is accustomed to think of as streaming through metallic wires, and magnetic fields as something that control the local direction of the needle of a compass.   But there are no metallic wires or compass needles out in space, so why should there be electric currents and magnetic fields there?   One has no feeling for the effects of currents which are distributed in space over large volumes or for the magnetic fields that they might generate.   We did not consider in the past that magnetic fields could produce electromagnetic phenomena as soon as they experienced a volume of gas in motion.   Another point is that the cosmic magnetic fields are often very weak.   Nevertheless, we have found that electromagnetic phenomena often become very important, so much so that they have a decisive influence on dynamics.   This is a result of the vast cosmic dimensions and the high electric conductivity of the hot ionized gases; ordinary hydrodynamics, i.e. the dynamics of ordinary uncharged fluids, does not suffice to describe what happens.   It has to be replaced by magnetohydrodynamics (MHD), a combination of electromagnetics and hydrodynamics [4.1–4.10].

We now know much more about such activities on the Sun.   There is lots going on there that we can observe—prominences, flares and Sun spots, for example—which all seem to be magnetohydrodynamic in nature.   Closer to home we have Earth's magnetic storms and auroras.   Interstellar space is electrically conductive and exposed to electric as well as magnetic fields.   And there is matter for them to act on, though not much, perhaps not more than one atom per cubic centimetre.

Once we see the cosmos as space supported by particles under the influence of whatever electric and magnetic fields happen to be acting there, we can begin to appreciate the importance of the theory Hannes Alfvén developed.   Recent experimental work—both on Earth and probing space—has given us an idea of the range of magnetic field interactions we have to consider.   How strong then are those magnetic fields? It is not surprising that the magnetic field intensities take very different values depending on where we observe them in space, or in the laboratory, for example

Interstellar medium	10^{-6} G
Galactic spiral arms	6×10^{-6} G
Solar corona	10^{-3} G
Solar wind	10^{-4} G
Sun spots	3×10^3 G
Earth's ionosphere	3×10^{-1} G
Earth's surface	5×10^{-1} G
Neutron star (pulsar)	$10^7 - 10^{12}$ G
Fusion experiments (JET)	34×10^3 G
Fusion reactor	60×10^3 G
Small motor (vacuum cleaner)	3×10^3 G

The field imposed by a small domestic motor such as the one that drives a vacuum cleaner is a few thousand gauss. Compared with that, the field in the interstellar medium is a tiny 10^{-6} gauss. At the other end of the spectrum is the field imposed on a plasma in the experiments to build a source of power based on the nuclear fusion, 4×10^4 G. But even this pales beside the fields Nature creates inside a neutron star which are 100 to 10 million times stronger. How is it possible to estimate the magnetic field strengths out in the cosmos, for example in spiral arms, and to determine the influence of the magnetic field on the dynamics of cosmic masses? These questions were addressed (in two separate publications) by S Chandrasekhar and E Fermi as common authors of both publications. In the first of these, two independent methods are described for estimating the magnetic field in the spiral arm in which we are located. The first method uses magnetohydrodynamic wave theory in terms of the Alfvén velocity to interpret the dispersion (of the order of $10°$) in the observed planes of polarization of the light from the distant stars. The second method is based on the requirement of equilibrium of the spiral arm with respect to lateral expansion and contraction. The condition for equilibrium is obtained by equating the gravitational pressure in the spiral arm to the sum of the material pressure and the pressure of the magnetic field (H^2). The first method gives $H = 7.2 \times 10^{-6}$ G, the second 6×10^{-6} G

[4.11]. The fundamental work on 'Magnetic fields in spiral arms' published in 1953 by S Chandrasekhar and E Fermi (*Astrophysical Journal* **118** 116) is reproduced in appendix 3.

The second paper discusses problems of gravitational stability in the presence of a magnetic field. In one example the authors consider the stability for transverse oscillations of an infinite cylinder of incompressible fluid, and for simplicity also infinitely high electrical conductivity. A uniform magnetic field is acting in the direction of the axis. It is found that the cylinder is unstable for all periodic deformations of the boundary with wavelengths exceeding a certain critical value, depending on the strength of the magnetic field. It is shown that the magnetic field has a stabilizing effect, and that for a cylinder of radius $R = 7.7 \times 10^{20}$ cm (250 parsecs) and mass density $= 2 \times 10^{-24}$ g cm^{-3}, a magnetic field in excess of 7×10^{-6} G effectively removes the instability. In another example it is shown that a fluid sphere with a uniform magnetic field inside and a dipole field outside is not a configuration of equilibrium. It will tend to become oblate by contracting in the direction of the magnetic field [4.12]. In spite of the simplifying assumptions these early (1953) analyses demonstrate the role of magnetic fields and the use of magnetohydrodynamics in cosmic applications [4.13].

Whatever the strength of the field, there is also the question of structure. In the cosmos and also in laboratory experiments, the magnetic fields, the currents and particle densities show filaments (as in the solar prominences) or granulations (as in the photosphere). Filamentation set up by the coupling of currents and magnetic fields in plasmas is noticed in the plasma tail of comets as well as in the tail of the Earth's magnetosphere [4.14].

The radiation from very powerful lasers makes small parts of the target very hot and very dense. As charged particles dash around they set up electric currents that generate structured patterns of magnetic fields as high as a hundred megagauss. The vortex is another structure we often see in magnetic fields, such as that in nebulas (e.g. the Crab) and galaxies (e.g. M83). Self-induced currents generate the fields, which form the structures emitting radiation to be captured by measurements. Relativistic electrons spiralling in the magnetic field structures generate synchrotron radiation spectra, as in the case of 'radio stars' observed in radioastronomy.

Vortices also organize themselves in auroras. The symmetries and patterns shown up in these observations provide a clue to the fine structure effects generated by magnetic fields acting in plasmas. Can Alfvén waves and magnetohydrodynamics help us to explain them? In my view the answer is in many cases, yes. In fact, the universe is really an immense laboratory for Alfvén waves. The results so far clearly indicate the outstanding role of the new type of wave motion that involves coupling between charged fluids and magnetic fields.

4.1.2 The challenge of space plasma exploration

Certain problems in space plasma physics, for example the acceleration of electrons in the solar wind and in the magnetosphere, demonstrate difficulties which are characteristic of geocosmological plasma research in general. Progress in the field of geocosmophysics has been astonishingly slow compared with that in other branches of modern physics such as elementary particle physics, atomic spectroscopy or nuclear physics. At first sight it seems, for example, quite remarkable that the mechanism of acceleration of the solar wind electrons and the auroral electrons is not known. Or that the phenomena at the solar surface and their connection with the internal generation of fusion power in the Sun are, in fact, still to be explained.

The difficulties of making detailed observations of particles and space plasmas by rockets or satellites should be rather obvious but are often underestimated. One important fact that has to be considered is that the medium to be observed is not controlled by the observer. As an example, try to relate observed data of certain phenomena, for example particle velocities or velocity distributions on a certain magnetic field line, to corresponding data observed at a later time and at a different point on the same field line. How can we be sure that it is the same field line? And how can we, therefore, be sure that we have the correct basic information to use for modelling phenomena, for example the acceleration of electrons? The difficulties in space plasma exploration are indeed very severe and require extreme skill in planning and realization of the measurements.

Acceleration under controlled conditions such as in huge laboratory accelerators as a rule operates satisfactorily from the outset. In space the acceleration mechanisms are not specified. The difficulties are there if one considers the phenomena on the solar surface, or on the outer portions of the solar corona from which the solar wind emanates, or if one is making observations of the magnetospheric bow shock or the magnetotail. One has to rely on measurements programmed and controlled at a remote distance with often sophisticated techniques. What a lot of difficulties to surmount and what a lot of interesting discoveries to make in the future! This is not to say that measurements and definitions of situations are not difficult in other branches of modern physics. The stochastic cooling of particles for example in large accelerators is certainly also something extremely sophisticated.

In the late 1950s when space research and space communication was just beginning I remember Niels Bohr, the father of atomic physics, expressing his admiration for the new satellites and the many possibilities they might offer. They encircled the Earth and could transmit a message instantly between any points on our planet. The master of electron motion and radiative transitions in atoms was impressed. It was at an evening gathering in the winter garden of his home, the honorary villa of the Carlsberg brewery in Copenhagen. I remember that it was a little to my surprise that such a strong appreciation of satellites was expressed by someone who had made such profound contributions himself to physics in general. Perhaps he could already envisage what a fascinating time there was to come . . . .

4.1.3   Cosmic plasma jets

An extraordinary space-based telescope has provided new possibilities for studying cosmic plasma jets. The NASA Hubble Space Telescope (HST) started operation early in 1990. The telescope, which has a primary lens 2.4 m in diameter, is carried by a satellite (length 13 m, diameter 4.3 m, weight 11.6 tons). As well as two spectrographs to analyse the radiation from celestial bodies, one for studying very faint objects, the other for more luminous objects but with higher resolution, the HST is also equipped with a Faint Object Camera (FOC). The FOC is capable of observing objects at least 30 times less bright than the best ground-based telescopes. Free from the influence of atmospheric disturbances the spectrometers of HST have a very wide wavelength resolution, from the extreme ultraviolet (115 nm) to the far infrared (1 cm), more than 3000 times wider than any ground-based telescope. The HST is ideally suited for studying extragalactic jets. The telescope's UV sensitivity allows it to clearly separate a jet from the stellar background light of its host galaxy. What is more, the FOC's high angular resolution is comparable to sub-arcsecond resolution achieved by large radio telescope arrays.

A remarkable photo of a 4000 light-year long jet of plasma emanating from the bright nucleus of the giant elliptical galaxy M87 was presented on 16 January 1992 at the 179th meeting of the American Astronomical Society in Atlanta, Georgia, USA (press release No STScI-PRC92-07). The ultraviolet light image was made with the European Space Agency's Faint Object Camera (see plate 5).

Allow me to quote the caption of the photo:

> M87 is a giant elliptical galaxy with an estimated mass of 300 billion suns. Located 52 million light-years away at the heart of the neighbouring Virgo cluster of galaxies, M87 is the nearest example of an active galactic nucleus with a bright optical jet. The jet appears as a string of knots within a widening cone extending out from the core of M87. The FOC image reveals unprecedented detail in these knots,

resolving some features as small as ten light-years across. According to one theory, the jet is most likely powered by a 3 billion solar mass black hole at the nucleus of M87. Magnetic fields generated within a spinning accretion disk surrounding the black hole spiral around the edge of the jet. The fields confine the jet to a long narrow tube of hot plasma and charged particles. High speed electrons and protons which are accelerated near the black hole race along the tube at nearly the speed of light. When electrons are caught up in the magnetic field they radiate in a process called *synchrotron radiation*. The Faint Object Camera image clearly resolves these localized sites of electron acceleration, which seem to trace out the spiral pattern of the otherwise invisible magnetic field lines. A large bright knot located midway along the jet shows where the blue jet disrupts violently and becomes more chaotic. Farther out from the core, the jet bends and dissipates as it rams into a wall of gas, invisible but present throughout the galaxy, which the jet has ploughed in front of itself.

Another extraordinary observation entitled: 'Hubble Space Telescope resolves braided galactic jet' (press release No STScI-PRC91-01) reveals new features of magnetoplasma interactions in galactic jets as described by the following presentation (see plate 6).

NASA's Hubble Space Telescope has provided a detailed view of a ten thousand light-year long jet of plasma which has been ejected from the core of a galaxy 270 million light-years away. Observations made with the European Space Agency's Faint Object Camera (FOC) reveal that the jet has an unusual braided structure, like a twisted pair of wires.

'This is the first time that such a structure has been seen in an optical jet', said F Duccio Macchetto, ESA's Principal Investigator on the FOC and Head of the Science Programs Division at the Space Telescope Science Institute. The results of this observation were presented at the meeting of the American Astronomical Society in Philadelphia, Pennsylvania on 15 January 1991.

The FOC image provides intriguing new details for under-standing how the core of an active galaxy generates such a narrow beam of energy and then propagates the jet across millions of light years, at velocities approaching the speed of light. The jet appears as a bright 'finger' extending in the northeast direction from radio galaxy 3 C 66 B. The FOC image has an angular resolution of 0.1 arcseconds which is 12 times better than previous ground-based optical images, and even three times better than high-resolution radio maps obtained with the Very Large Array radio telescope at Socorro, New Mexico. In addition to the unique double-stranded structure, the FOC image reveals filaments, bright knots, kinks and other complex features never before seen in an optical jet. Many of these features overlay the radio structure of the jet.

The jet was observed on 28 August 1990 as part of an early programme to assess the optical performance of the HST. According to Macchetto 'The HST is a uniquely important instrument for studying synchrotron jets'. Image reconstruction techniques applied to the data bring out additional detail.

In addition to the HST's extraordinary resolution its ultravi-olet sensitivity is ideal for studying optical jets because they are relatively bright in the UV compared with their host galaxies. Once the image of the host galaxy is subtracted through computer pro-cessing, details in the jet can be traced all the way to the galactic core.

The bluish, highly polarized light of the jet in galaxy 3 C 66 B is produced by electrons which are spiralling along magnetic fields at velocities approaching the speed of light. The light and radio emissions produced by the electrons are synchrotron radiation—so named because similar radiation is observed in particle accelerator machines. But what is the 'machine' which is the powerhouse behind the jet in 3 C 66 B? The favoured mechanism is a super-massive black hole which may lie at the core of the galaxy. Stars, dust and gas swirl deep into the hole's intense gravitational field along a broad flattened accretion disc. The hot plasma in the spinning disc creates powerful electric currents which in turn generate twisted magnetic fields which align with the black hole's spin axis. The black hole's spin axis is also an escape route for the high-speed electrons. As the electrons spiral outward along magnetic field lines, they lose energy in proportion to their frequency and the strength of the magnetic field. In a timespan of only a few hundred years the electrons responsible for the optical emission lose much of their energy. However, the electrons which produce radio emission can survive in the same magnetic field for tens of thousands of years. Hence most galactic jets which extend for thousands of light-years are detected in radio wavelengths. In only a few cases have optical counterparts been observed.

Nevertheless, the long optical jet in 3 C 66 B presents a mystery for astronomers. How do the electrons remain energetic enough to radiate visible light throughout their 10 000 year journey along the jet? One possibility is that the electrons are boosted back to higher energy levels as they move along the jet, perhaps by instabilities at the edge of the plasma flow. Another possibility, which is suggested by the jet's braided appearance, is that the electrons speed along a channel which has a much lower magnetic field strength and hence lower energy loss.

Two sharp bends and kinks in the strand (3000 and 8000 light-years out from the nucleus) are also hard to explain. They may indicate that the galaxy's 'central engine' does not release energy at a steady rate but rather 'hiccoughs' or fluctuates in output. The kinks may also be produced by a complex magnetic field structure along the jet, or collisions with dense regions in interstellar gas.

3 C 66 B is only the second galaxy with an optical jet which has been observed by the HST (the previous optical jet studied is in the galaxy PKS 0521-36). Both observations show a close match between the optical and radio features in the jets in each respective galaxy. However, there are significant differences between the structure of the jets of these two galaxies. This suggests that different mechanisms are at work in transporting material from the galactic core.

Further observation with the HST will provide fundamental new information about the nature of galactic jets, in particular how energy is transported along the jet and the role of magnetic fields in channelling material from the core of a galaxy to intergalactic space.

Attempts have been made to explain the formation of extragalactic jets by Alfvén waves. The generation of Alfvén waves is assumed to originate from the twisting and annihilation of magnetic fields, and the damping of the waves is provided by nonlinear, surface and turbulent damping. The magnetic field configurations in extragalactic jets are derived from polarization observations, which imply that the magnetic fields in the powerful jets are longitudinal. The model for the jet outflow is based on knowledge of the magnetic structure of the Sun and experience from studies of the solar wind. The jet flow velocity u_j is estimated indirectly from observations, the velocities ranging from 1000 km s^{-1} to the velocity of light. The characteristic period of the Alfvén waves is from 10^4 s to 10^9 s which is in agreement with fluctuation periods for the radiation continuum of quasars and active galactic nuclei. The conclusion is that Alfvén waves with their presumed damping mechanisms can be important for the acceleration of extragalactic jets.

4.1.4 The all-pervasive Alfvén wave

Hannes Alfvén himself regarded the space age as being a revolution in science, comparable to the introduction of the telescope by Galileo Galilei. He pointed out the ability of spacecraft to serve as platforms for instruments to observe remote objects and their advantages *vis-à-vis* ground-based telescopes for optical observations. It seems that we are still only at the beginning of the space age and that we can expect a continued rich harvest of new results in the future.

Alfvén waves are of fundamental interest for geocosmophysics, and the field of Alfvén waves is an active area of research in all branches of plasma physics more than half a century after their discovery. As early as 1937 Hannes Alfvén suggested the existence of a galactic magnetic field. He was led to this proposal in an attempt to understand where cosmic rays obtain their energy from [4.15].

Apart from being considered as the mechanism responsible for the acceleration of particles to high energies in cosmic rays, Alfvén waves have been proposed as the source of momentum for the solar wind. Their presence has been confirmed by the interpretation of experiments by the spacecraft Mariner 2 in 1968. Particle scattering by Alfvén waves also helps to maintain the isotropy of cosmic rays in space.

How is it possible to identify Alfvén waves in space and to confirm the effect that they may have? A characteristic feature of Alfvén waves is that the perturbation velocity and perturbation magnetic field are parallel and proportional to each other. This is true not only for continuous, i.e. periodic, waves but also for bursts. This simple relationship presumes that the medium may be considered as incompressible, which is in general a fairly good approximation. For the solar wind one can show that a necessary condition is that the Alfvén velocity v_A is very much less than v_S, where v_S is the speed of sound.

The measurements acquired by Mariner 2 show strong correlations between the velocity fluctuations and the magnetic field variations. Twenty-four hour observations of magnetic field and plasma data demonstrate the presence of nearly pure Alfvén waves.

The *generation* of Alfvén waves presumes changes in magnetic fields which, in fact, occur frequently in space. To *transfer* energy from the Alfvén waves and give momentum to particles, one has to rely on processes that *damp* the waves. Acceleration of electrons, heating of plasmas, or current drive, i.e. maintaining plasma current, may be achieved in this way. There are various processes that may cause changes in the magnetic field and thereby provide conditions for generation of Alfvén waves, for example the annihilation of twisted magnetic fields near a stellar surface. Nonlinear and turbulent damping mechanisms could transfer energy from the Alfvén waves, for example in the solar wind [4.16].

Alfvén waves have, furthermore, been considered as candidates for heating the solar corona and in modelling of an abrupt transition jump in temperature in the outer parts of the solar corona. The presence of Alfvén waves in a cometary environment was identified by observations in 1985–6 (see section 4.7). Alfvén waves have been used in describing phenomena in planetary magnetospheres, e.g. magnetosphere–ionosphere coupling, micropulsations, aurora formation, etc. Alfvén waves also play an important role in the formation of quasar clouds and of extragalactic jets. Alfvén waves are better understood now thanks to laboratory experiments that use them in various ways to heat plasmas. They may also be used to maintain the plasma current in fusion plasma devices. The results obtained so far clearly indicate the outstanding role of the coupled systems of magnetic fields and electric currents in the cosmos as well as for fusion plasmas on Earth.

4.1.5   The dynamic face of the Sun

Our Sun as we know it from daily life is continuously shining with a friendly yellow-reddish colour on its face. It has most likely looked the same and remained the same size for billions of years. But seen in more detail by a telescope, particularly with a high-resolution satellite-based spectroscopic instrument, things may look different. The Sun bubbles all over the surface and occasionally very dynamic phenomena may occur.

Looking at these is like watching a TV camera survey a football stadium where a match is going on. Expectations have excited the atmosphere and waves of hope and dissolution wander over the stadium. And suddenly, after some dramatic preparations, it happens: a goal for the home team! Supporters suddenly throw hats and newspapers and fireworks high in the air. Applause is accompanied by roars and howls which can be heard all over the city. The stored energy has found a way out and continues to excite the spectators as well as the players for some time.

We can take another analogy from a tennis tournament at Wimbledon or Roland Garros. Mostly, conditions are calm. But once in a while the players show signs of happiness or despair, throwing their rackets and balls high in the air and shouting at the umpire for unexpected decisions. Liberation of internal energy can take many forms!

As we shall see, this is just what is characteristic of the phenomena that may occur on the solar surface driven by the energy liberated from the interior of the Sun. Enormous gas clouds expand outwards, carrying with them magnetic fields and currents in often twisted and filamentary structures. It is these that give rise to phenomena on the solar surface, such as Sun spots, solar flares and the prominences which exhibit gigantic arcs far out from the surface [4.17].

Telescopes show us many characteristic motions and detailed traces of internal activity that occur on the solar surface. Even with the strongest optical telescopes we cannot easily see such fine details on other stars. So the Sun provides a unique opportunity to study interesting phenomena which presumably also take place on other stars in the universe. Thanks to instruments on the new satellites we can see the structures giving rise to the phenomena on the solar surface. Now we can embark on new investigations of basic features of the dynamic face of the Sun. We look at how heat flows through matter and what is generating the magnetic fields inside the Sun. We also have results concerning how energy is released and transported in the Sun's outer layers using the mechanisms of radiation, mass motion, heating and particle acceleration.

It is important to link together such experiences to enable us to make comparisons between results from laboratory and natural plasmas with regard to magnetic reconnection, i.e. self-induced magnetohydrodynamics, joining of magnetic field lines and, furthermore, acceleration processes. The beautiful structures of filaments and vortices we see today in prominences ejected from the solar surface demonstrate several types of plasma instability caused by magnetic field–plasma current interaction. These are already well known from laboratory experiments and are called, for example, kink, sausage and filamentation instabilities. They may be regarded as external traces of what is going on under the solar surface, a domain to be explored by future research.

Galilei first noticed dark spots on the solar surface in 1610. His observations made with telescopes, which were still a novelty, were regarded with great scepticism. The Sun was considered as a God which should have no imperfections. Repeated observations confirmed, however, that the spots existed and that they occurred in groups of two or more. We now know that they are regions of gas on the surface of the Sun which are cooler than the rest of the surface. They have a cycle of repetition of about 11 years. The spots consist of a central region, the umbra, plus the outer grey region, the penumbra, which shows a filamentary structure which extends into the granular structure of the solar surface.

The diameter of the Sun is more than 100 times greater than the diameter of the Earth and Sun spots can easily be larger than our continents on Earth. Strong magnetic fields, of several thousands of gauss occur in the spots, whereas the general magnetic field on the Sun has a strength of about one or two gauss, the magnetic field on the surface of the Earth being about half a gauss. Neighbouring spots have opposite magnetic polarities. No-one knows how these strong active fields are generated. One theory is that different layers in the Sun rotate at different speeds, producing vortices in the solar interior. Maybe vortex motion carries the magnetic fields to the surface of the Sun, where they can be observed by their influence on atomic spectral lines or by their interactions with ionized gas.

4.1.6 Solar flares and prominences

Sun spots are violent, turbulent places. For example, strong eruptions of the Sun produce *solar flares*, bright flames from the solar surface. Energetic particles from the eruption enter the Sun's atmosphere where some of them interact to generate radiation in the optical, UV, x-ray and gamma ray parts of the spectrum. The rest of the fast particles as well as clouds of solar plasma are ejected out into interstellar space. Some reach the Earth, where they can cause magnetic storms in the Earth's field. Solar flares occur abruptly and last as a rule for 15–20 minutes, sometimes an hour or two. Sometimes they take the form of flare surges,which can reach more than one million kilometres in height, carrying plasma and magnetic fields. They are ejected with velocities of the order of 1000 kilometres per second. When they reach the Earth they generate spectacular auroras and can interfere with radio communications.

Prominences are more long-lived gigantic gaseous eruptions, which may reach several hundreds or thousands of kilometres above the solar surface and stay there for long periods, sometimes many hours. It may well be the influence of magnetic fields that maintains them before they return to the solar surface. The prominences also follow the 11-year solar cycle.

The first publication mentioning a solar prominence had the title 'Observation of a solar eclipse in Göteborg, Sweden May 2nd 1733'. It was by Birger Vassenius and was published in 1733 *Proceedings of the Royal Society* no 429 pp134–5. Here is a translated extract from the Latin:

> During the time when the whole Sun was covered, I saw, apart from a great deal of the spots on the surface, the atmosphere of the moon in a telescope of about 21 foot size: and I saw it on the west side of the moon, under the maximum position, somewhat more luminous; though without the irregularity and unsmoothness of the light-beams, which entered the eyes of a spectator without tube. Meriting not only admiration, but also a notice from the illustrious Royal Academy, seemed to be some reddish spots which were noted exterior to the moon periphery, three or four in number; among them one larger than the others, almost in the middle between south and west, as far as it was possible to estimate. It was composed of something like three parts or smaller parallel minor clouds of unequal extensions somewhat oblique with regard to the periphery of the moon. Since I was caught by admiration of the phenomenon, I gave a friend, whose eyes were very sharp, an opportunity to observe. But when it turned out that he, who was not acquainted with a tube, could not even find the moon, I was

happy to continue myself for 40 or more seconds to regard the same spot, or if one so desires, cloud, which was unchanged and stayed in its old position, near the moon's periphery, notwithstanding any suspicion of fault with the tube or the eye . . ..

The French astronomer Camille Flammarion portrayed prominences in his important work *Astronomie Populaire*, published in 1880, which greatly stimulated the general interest in astronomy. It may be noted that even Hannes Alfvén became interested in astronomy from this work. The book contains beautiful etchings in colour (Bordeaux red) of magnificent prominences observed in 1872 at Harvard College Observatory, USA with detailed descriptions and discussions of the events (see plates 7 and 8). These plates show two magnificent prominences, the first one (plate 7) on 29 April 1872, the second one (plate 8) on 15 April of the same year and the same hour, 10 o'clock in the morning. The scale of the prominences is such that the horizontal distance amounts to 200 000 km. To give some feeling for the size we may note that this distance is about that which a modern car covers in a lifetime.

To justify the natural beauty of the prominences it is necessary to reproduce those solar flares in colour. That is why the two excellent chromolithographies from Harvard College Observatory, USA, are reproduced here. There is, however, something that even the colour pictures could not reproduce, namely the impression of life and rapid changes in the pattern that is characteristic of those phenomena. It helps to think of violent flames when imagining what happens in the pictures! They really demonstrate the beauty of plasmas, and of the magnetohydrodynamics which lies behind it. In these pictures, which are artist's impressions rather than photographic images, one can see filaments and blobs of gases forming. These indicate plasma instabilities and the presence of magnetic fields, providing curved motions of the plasma. A great variety of black and white images exist which show time sequences and indicate the various shapes that have been observed in prominences, from flame-like outbursts to cloud-like shapes which sometimes approach the pattern of a woven tissue. These demonstrate the fascinating variety of ways in which energy can tunnel out of the gravitational field that confines it!

On 3 April 1873 a luminous hydrogen cloud of enormous height was observed outside the limb of the Sun. It seemed like a cirrus cloud, light and filamentary; the entanglement being very difficult to grasp and changing from one instant to another. From being straight and diffuse at the beginning it ramified rapidly into branches of a twisted spiral nature after 25 minutes to extend to a height of 322 000 kilometres, i.e. about one-quarter of the diameter of the Sun, with an average speed of 105 kilometres per second.

Once again, it may be tempting to return to art. Vincent Van Gogh in his paintings emphasized sunflowers and cypresses as two types of object that he considered as contrasts or complementary to each other. Sunflowers represented open structures catching energy from the Sun to feed the plant, whereas cypresses had the form of confined, closed structures keeping their identity as they evolved. Is there not an obvious similarity between Sun spots and sunflowers as there is also between cypresses and prominences? His artist's intuition led him to analyse objects on the artistic scene which happen to closely resemble objects in modern solar physics research.

The shape and structure of the outbursts are indeed, as can be seen from the plates, strikingly akin to the cypresses in the Vincent Van Gogh paintings 'Starry night' and 'Path with cypresses and stars' (see also the smaller cypresses in the second of these paintings) (plates 2 and 3). The pictures in Camille Flammarion's *Popular Astronomy* were published in 1880 when Vincent Van Gogh was 23 years old, whereas the paintings were made in 1889–90, the last years of his life. It is not entirely inconceivable that Van Gogh was even inspired by the unique popular astronomy book as was later, at a very early age, around 1920, the creator of magnetohydrodynamics Hannes Alfvén. If so, what a coincidence!

We must hope that future research will elucidate the connection between Sun spots, solar flares, prominences and vortex structures, and relate them to the energy released at the centre of the Sun by fusion reactions. Great advances are already being made in solar observations with the new satellite-based instruments.

4.1.7 Comets

Comets are spectacular objects in the sky characterized by a head of nebulous gas surrounding the nucleus and a tail of luminous gas which can be several millions of kilometres long.

The nucleus of a comet consists of small frozen particles of gas and has dimensions of several kilometres. It probably contains pieces of stone, sand and dust.

Comets are part of the solar system and they often move in strongly elongated elliptical orbits. When a comet approaches the Sun the material of the comet gets hot. Frozen gases, water, carbon dioxide and ammonia evaporate and evolve to form the gaseous head of the comet which is lit by intense short-wavelength radiation from the Sun. Spectra of emitted light from the comet reveal the presence of hydrogen, nitrogen, carbon and oxygen, and CH, OH and CN radicals, ionized as well as neutral. Intense particle radiation from the Sun, i.e. protons and electrons of the solar wind, also strikes the comet and causes an ionized comet tail, which is directed away from the Sun. Experiments have been done to stimulate comet tails in which plasma has been ejected from rockets at a height of about 2000 km.

According to Alfvén the interstellar magnetic field from the solar wind plays an essential role in the formation of comet tails. The magnetic field accounts for structures often observed in the tails. Spectroscopic analyses show that electrically charged particles are present in the tail. The plasma consists of CO^+, CO_2^+, CH^-, N_2^+, OH^+, H_2O^+ and free electrons. CO^+ and N_2^+ are the most long-lived (10^6 s) and can therefore exist at great distances from the head of the comet. The velocity of the gas particles can be 10 km s^{-1} near the gas head, but can be 100–1000 km s^{-1} far out in the tail.

In 1985–6 the first observations of Alfvén waves in a cometary environment were made by satellite observations of comet Giocobini–Zinner and comet Halley. The generation of Alfvén waves near comets had been discussed in theory long before the observations, which thus confirmed the predictions of the theory. It seems that the physics of Alfvén wave generation in cometary environments is fairly well understood by now, unlike the situation in the solar wind. It may, in fact, be that the results of cometary investigations might help us obtain a better insight into the corresponding solar wind problem. Let us therefore take a look at what is behind the presence of Alfvén waves in comet tails.

In interplanetary space a certain number of neutral particles may always exist. They could be of interstellar, planetary or cometary origin. When these particles come close to the Sun they may be ionized by solar radiation. From neutrons and helium the radiation will produce protons, alpha particles and free electrons. As soon as these are produced they immediately start to interact with the fast-moving solar wind. The result will be a number of wave modes, electromagnetic as well as plasma waves. Numerical simulations have shown that all unstable modes are insignificant except the Alfvén mode, which can increase to high levels, whereas the others become stabilized due to the low density of the newly produced ions as compared with the density of solar wind particles. As a result of the interaction the generated Alfvén waves can also enable the solar wind to pick up the created ions. Observations have shown that comets continuously eject neutral particles and gas which become ionized immediately, including heavy ions like CO^+ etc. It is fascinating, as least for a theoretician, to think of how the all-pervasive Alfvén waves lie behind the spectacular phenomena of comet tails.

A new area in cometary research is the observation of x-rays from comets. Plasma data obtained by satellites from comet Halley, for example, suggested that comets should emit x-rays. An energetic electron population with energies up to several keV was observed. Strong plasma wave turbulence in the lower hybrid frequency range ω_{lh}, equal to the square root of the product of the electron (ω_{ce}) and ion (ω_{ci}) cyclotron resonance frequencies, was simultaneously noted. Computer simulation studies had shown that lower hybrid waves were produced when comets interacted with the solar wind. Such waves seemed to be active at accelerating electrons parallel to magnetic field lines, to energies required for x-ray generation.

Recently, systematic observations were made of x-rays from comet Hyakutake. They gave the astonishing result that the observed signals turned out to be more than a hundred times stronger than the predictions. Subsequent analyses and simulations [4.18] have, however, recently confirmed the observed intensities. The studies were based on a lower hybrid instability for the interpenetrating cometary and solar wind ion gases.

X-ray production should depend on the solar wind. Observations of x-rays from comet Hyakutake indeed showed time variations of the x-ray fluxes. Comets may therefore be used as remote probes of the solar wind.

Statistics about comets have been collected for many hundreds of years. Since the days of the invention of the telescope, experience shows that only about one-tenth of the total number of astronomical events have been observed by the naked eye. The number of reported comets before Galileo should therefore be multiplied by at least a factor of ten to yield a comparable total as related to observations with telescopes in later years. From the years since AD 1 the effective number of observed comets would then amount to several thousands in total, or several each year.

When the question was raised 'How many comets are there in the sky?' the astronomer Johannes Kepler answered 'As many as there are fishes in the oceans', which was not an exaggeration. According to modern estimates the solar system is surrounded by an extended cloud that includes 100 billion comets. Comets could be useful probes for diagnostics of the interstellar medium, for solar particle ejections and, in the case of impact, for observations on planetary crusts. They should therefore continue to be challenging objects for continued astrophysical research.

Outstanding achievements in the history of cometary observations have been made using the Hubble Space Telescope. It imaged the 'String of pearls comet' (see plate 9), a train of 21 icy fragments across 710 thousand miles (1 million km) of space, or three times the distance between the Earth and the Moon. Hubble's high-resolution image taken with the Field and Planetary Camera showed that the comet's nucleus was much smaller than originally estimated from observations with ground-based telescopes. The Hubble observations showed that the nucleus was probably less than 3 miles (5 km) across, as opposed to earlier estimates of 9 miles (14 km). The pictures discussed above were taken a year before the expected impact of the multiple comet on the planet Jupiter. On the occasion of the impact in July 1994 Hubble pictures taken with its Planetary Camera showed eight impact sites (see plate 9).

The comet, named Shoemaker–Levy 9 after the discoverers, attracted attention from telescopes around the world when the collision with Jupiter occurred. The extraordinary spectacle continued for several days as fragments of the comet caused a series of giant explosions including fireballs over a 1000 miles across (press release No STScI-PRC94-26a).

In July 1995 two American astronomers, Alan Hale and Thomas Bopp, simultaneously but independently discovered a new comet which accordingly has been named Hale–Bopp (see plate 10). The luminous head of the comet extends into two magnificent tails about 3 million kilometres long. The white tail, which is the more dominant and broader of the two, contains dust which reflects the solar light. The second one, which is blue and has a filamentary structure, is composed of ionized gas from the solar wind and is in the direction opposite to the Sun. It is produced by the solar wind of plasma emitted from the Sun. Hale–Bopp produces 250 tons of water per second. Interestingly enough, the comet exhibits all four states of matter: solid state (rock and ice), liquid (water), gas and plasma. The diameter of the head of Hale–Bopp is estimated to be about 40 km. Since its discovery, astronomers have had the opportunity to study the comet and determine the chemical structure of its dust, rock and ice, the composition of which is supposed not to have changed much since the birth of the solar system about 4.5 billion years ago [4.19, 4.20].

By April 1997 some 33 molecules had been detected, among them SO_2 for the first time in a comet. Meanwhile European astronomers discovered that Hale–Bopp also has a third tail consisting of sodium (Na). The third tail, which contains sodium in the form of uncharged atoms, is emitted in a direction close to the blue ion tail. On 22 March 1997 Hale–Bopp reached its closest distance to the Earth, about 200 million kilometres. It then had a velocity of 44 kilometres per second. Remember that the distance between the Sun and the Earth is about 150 million kilometres, i.e. less than the closest approach of Hale–Bopp. In March and April 1997 comet Hale–Bopp was visible every night. The comet's tail could be seen clearly with the naked eye. Looking at the comet with binoculars, as I happened to do against a completely clear sky on the French Atlantic coast, was an extraordinary experience of a celestial event. Comet Hale–Bopp will not return for another 2380 years.

The dust lost by comets in periodic motion about the Sun is distributed along the orbit of the comet. When the Earth in its motion around the Sun happens to cross the orbit of a comet it attracts the dust particles, which in the Earth's atmosphere induce bright sources of light. The particles, which may be no larger than a millimetre across, enter the Earth's atmosphere at high velocity, for example 10 kilometres per second, and become split into atoms and ionized in collisions with air molecules in the direction of the orbit. This occurs at a height of 80–120 km above the Earth. A *meteor* trail is formed. On certain occasions whole bundles of meteors fall from certain directions in the sky, as in the nights around 12 August every year on the event of the Perseid showers. The meteor trails or 'falling stars' of these showers have also been named 'Saint Laurent's tears'. Occasionally, the luminosity of the ionized meteor trails, or plasmas, is comparable to that of the full Moon. The plasma in the meteor trails can be studied by means of pulses of electromagnetic waves using radars. And, when you see a meteor, don't forget to make a wish, but keep your wish secret!

Is it the midnight's comb
which makes them fall
the stars of spring time?

(Natsume Sôseki 1867–1916, free translation from the Japanese
Kasamakura).

4.2   The Sun–Earth interplay

4.2.1   The solar wind

During periods of strong solar activity intense bursts of plasma and
electromagnetic radiation occur on the solar surface. The release
of energy can be enormous. A single event can liberate energy
comparable in amount to the total energy consumed by humans
on Earth since the beginning of time. Consequences of such
phenomena are experienced in the magnetosphere surrounding
the Earth. Secondary effects can be observed, for example, in
auroral activity. They can also be noted as perturbations in radio
communications, or in signals guiding spacecraft and satellite
performance. There is, however, also a permanent continuous
flow of particles in all directions out of the Sun, the solar wind,
which consists essentially of protons and electrons. This wind can
be influenced by solar magnetic fields, which occur in connection
with Sun spots. The unperturbed wind has an average velocity
of about 430 km s^{-1} when it reaches the region of the Earth, a
value which could increase to 1200 km s^{-1} when solar eruptions
occur. At a speed of 430 km s^{-1} the solar wind reaches the
Earth in about four days. The solar wind has a high electrical
conductivity. Accordingly, changes in magnetic fields will be
compensated by new magnetic fields accompanying induced
electric currents. Solar wind electrons have thermal velocities
which are considerably higher than their drift velocity, whereas
the opposite is true for the ions. All stars of the same type as the
Sun are believed to put flows of matter out into interstellar space,
which should accordingly be filled with intercrossing plasmas
from stellar winds.

The solar wind represents a plasma link between the Sun and the Earth. It was not until a century ago that it was pointed out more directly that beside visible radiation the Sun emitted some kind of particle stream, which prevailed even in the absence of enhanced solar activity. In 1896 the Norwegian physicist Kristian Birkeland suggested that the aurora borealis might be caused by electrically charged particles from the Sun which entered and were guided by the Earth's magnetic field to regions near the poles. Electric discharge tubes had at that time been recently developed, and Birkeland thought that the light from them looked similar to the aurora and that the phenomena might be of the same nature. More conclusive evidence of the solar wind came much later, in the early 1950s, from studies of comets by the German scientist Ludwig Biermann. In his studies he found that the pressure of electromagnetic radiation from the Sun could not be sufficient to explain that the tails of comets were always directed away from the Sun. Biermann concluded that streams of charged particles from the Sun were responsible for this effect and, furthermore, that the solar wind was blowing continuously. The early ideas about particle emissions from the Sun were also confirmed by spacecraft and satellites in the late 1950s and in the 1960s.

There remain many puzzling questions of a physical nature related to the interaction of the solar wind electrons and magnetic fields in the magnetosphere. Electric fields are also believed to play an interesting role. During their long journey from the Sun to the surroundings of the Earth solar wind particles undergo many adventures. When the solar wind electrons leave the outer parts of the solar corona they are believed to have acquired high, even relativistic, velocities. The acceleration mechanism is simply not known, although various proposals have been made. In passing the bow-shock the electrons seem to slow down their energy considerably, whereas for some reason they become accelerated again before entering the Van Allen radiation belts where they can have energies of some MeV and even higher. Spiralling in the Earth's magnetic field they contribute to the radiation belts which encircle the Earth at distances of 1.5 and 6 times the Earth's radius.

When high-energy relativistic electrons penetrate interstellar space and encounter the magnetic fields of interstellar objects they radiate synchrotron radiation and take part in the radiation from radio sources in outer space. They form the basis of an important, relatively new branch of astronomy, namely radioastronomy, which during the last 50 years has considerably increased our knowledge of the universe, shedding light on radiation processes and chemical abundances of elements at very remote distances in the cosmos, using Earth-based telescopes for the observations.

There are many electromagnetic processes that radioastronomy and fusion plasma sciences have in common. For example, when relativistic spiralling electrons radiate in the presence of a magnetic field they also lose energy which limits the generation of synchrotron radiation. These phenomena are present also in fusion plasmas, where losses by synchrotron radiation play a role in magnetic confinement experiments. In laser fusion experiments, where self-generated megagauss magnetic fields could be present, the strong electromagnetic laser fields could even drive the laser pellet electrons to relativistic speeds of electron oscillation. Relativistic mass corrections then become so important that further increase in the laser intensity could become inefficient.

In considering solar wind problems, the following basic properties should be remembered. In the solar wind, collisions are so rare that they can be neglected. Interaction of the solar wind with electromagnetic fields and waves thus becomes particularly interesting. The energy density of the ions is much greater than that of the magnetic field and also much greater than that of the electrons. Thus the ions, i.e. the protons, dominate the motion. As we shall see, however, the influence of electric fields set up by space-charge effects could be important. An interesting observation is that the solar wind can flow at an angle with respect to the interplanetary magnetic field without being deflected. How can this be possible? The answer lies in the influence of the space-charge, or more precisely in the strong electric fields set up when positive and negative electric charges tend to separate from each other due to the influence of a magnetic field, which would normally deviate the positive and negative charges in opposite directions if they were left alone without mutual electric field interaction. But this is, in fact, not the case and space-charge effects can prevent the separation. It seems that transverse inhomogeneities of the plasma with regard to the direction of the solar wind would increase the role of the space-charge, and they would even be necessary to prevent the opposite charge particle deviations of the orbits if the magnetic field were homogeneous in space. Filamentations in the plasma flow would thus tend to keep the protons and electrons together over long distances in the interplanetary magnetic field. It is therefore interesting to notice that in fact plasma filamentations seem to be common in cosmic plasmas.

4.2.2 The Earth's magnetosphere

The term magnetosphere was coined in 1959 by Tom Gould, an American scientist and pioneer in space and plasma astrophysics. It was introduced to give reference to the region in space where electrodynamic forces dominate the motion of plasmas in the presence of a planet's magnetic field.

Our new perception of the Earth's magnetosphere emanates from the interaction of the solar wind with the magnetic field set up by the Earth. The solar wind is actually driving the whole structure. The result is a complex of currents, of trapped particles forming confined plasmas, of shock fronts, plasma sheets, instabilities and turbulence, and of a magnetospheric tail stretching out at the back of the whole structure. There are currents in the ionosphere flowing into the upper atmosphere and currents linking the magnetosphere with the ionosphere in the polar regions which play an interesting role in the behaviour of auroras. It has been observed recently that in the magnetosphere the solar wind seems to drive surface waves on the transition region between the space plasma and free space like the wind drives waves on a water surface!

The most striking evidence of the solar wind is perhaps the change in the picture of the magnetic field surrounding the Earth from the idea one had only 50 years ago. It was William Gilbert who, in the year 1600, initiated the science of geomagnetism and noted that 'The Earth globe itself is a great magnet'. It has been known for 200 years that the magnetic field on the surface of the Earth can be approximated with high precision by the field of an imaginary bar magnet inside the Earth. Before the knowledge of any external influence on the magnetic field outside the Earth it was believed that a dipole structure like that from a bar magnet continued even outside the Earth into interstellar space.

Modern space research has resulted in striking evidence of a dramatic change in the picture of the magnetic structure around our planet. A beautiful demonstration of the effects of a streaming plasma, the solar wind, on the formation of a self-consistent magnetoplasma structure on a global scale includes many internal peculiarities. Some of these can be compared with what happens in large-scale fusion plasma experiments. As an example it may be mentioned that the problem of magnetic reconnection in plasmas where collisions can be neglected was first considered in connection with space plasma processes occurring in the Earth's magnetotail. Later on the same problem became of interest in research into fusion laboratory plasmas. It was found to have application to the so-called saw-tooth crash of the central temperature of a tokamak plasma. High-amplitude increasing oscillations become quenched suddenly on a short time-scale compared to the average electron–ion collision time due to practically instantaneous magnetic reconnection.

Early satellites (Pioneer I (1958), Pioneer V (1960)) observing the magnetic field around the Earth recorded a sudden decrease of the magnetic field at about 14 Earth radii. It was confirmed by further satellite measurements that the abrupt transition in the magnetic field strength to a very weak interplanetary field occurred at the position of a shock front, the bow-shock, which forms the outer limit of a region of an irregular and fluctuating magnetic field, called the magnetosheath, which contains a plasma in a turbulent subsonic state in contrast to the supersonic solar wind plasma (Mach number 8) outside the bow-shock.

The origin of the bow-shock is the necessary deflection of the solar wind in front of the Earth's magnetic field. The process of forming the bow-shock is believed to involve the interaction of waves of different types, increasing the temperature of the plasma, which forms a blanket or magnetosheath. Bow-shocks occur when a fluid or a plasma passes unmagnetized or magnetized bodies. They can be produced and studied in wind-tunnel experiments. The creation of the bow-shock depends on the relative motion of the fluid and the body. Comets such as comet Halley have bow-shocks due to their motion in the interplanetary medium. Comet Hale–Bopp exhibits two bow-shocks for reasons that so far seem to be a puzzle. The flow of the solar wind through cometary bow-shocks offers interesting opportunities for future research, as does the physics behind the bow-shock of the Earth's magnetosphere. Inside the bow-shock of the Earth's magnetosphere there is a region of plasma named the magnetopause, and furthermore the magnetotail, a stretched-out plasma behind the Earth, with the interesting property that the magnetic field has opposite directions above and below the equatorial plane.

The Earth's magnetic field forms a closed magnetic structure and gives rise to confinement of high-energy particles consisting of protons and electrons in the Van Allen radiation belts, named after their discoverer. The first observations of these radiation belts were made by means of the satellites Explorer I, II, IV and the lunar probe Pioneer III. The discovery of the Van Allen belts marked one of the great early successes of space science.

From the late 1950s modern space technology thus started to provide the means for observations of details of the magnetosphere such as magnetic and electric structures and of particle density and velocity distributions. As a result our modern view of the magnetosphere emerged—the strongly deformed overall magnetic structure and the complexity of the magnetoplasma and the current paths. Even more interesting from a scientific point of view is the fact that the experiments demonstrated the important influence of electrodynamics and plasma physics on the behaviour of matter and radiation in the Earth's environment.

In this context it seems interesting to remember the fact that as a rule gravitationally confined plasmas (the stars) emit visible radiation, which one can see in optical telescopes or with the naked eye, whereas magnetically confined plasmas emit radio and/or microwaves, which cannot be seen visually, but can be observed by rockets or satellites outside our atmosphere or by radiotelescopes operating in 'windows' of our atmosphere where the waves penetrate. This may be considered as a scientific reason for why electromagnetic knowledge about the universe advanced so slowly up until the space age.

The results obtained in the space age confirm and extend the pioneering early work of many scientists, e.g. Carl Størmer in Norway, at the beginning of the century, Hannes Alfvén in Sweden and Van Allen in the USA, who considered the motion and trapping of charged particles in the Earth's magnetic field. Today magnetospheres are known to exist on other planets in our solar system, such as Saturn, Jupiter, Uranus and Neptune, but not on Venus or Mars. In observations of galaxies one has found magnetospheres on a galactic scale, and a whole galaxy that shows a deformed structure with a tail due to the presence of an interstellar wind. Associated phenomena have also been studied by numerical simulations on computers and these studies have confirmed the observations.

4.2.3   The ionosphere

An ionized layer surrounds the Earth at heights above 100 km. It is generated by the ultraviolet (UV) radiation from the Sun, which ionizes gas in the atmosphere, which then becomes a plasma. A plasma equilibrium state is established by the balance between UV radiation, which tends to increase the number of free electrons, and the effect of recombination between free electrons and ions, which causes a loss of free electrons proportional to the square of the density of free electrons (assuming equal densities of free electrons and ions). The equilibrium and dynamics of the ionosphere in the presence of variations in the radiation, such as occur daily or on a shorter time scale, for example during a solar eclipse, are therefore examples of nonlinear phenomena. The ionosphere reflects radio waves in the broadcast frequency domain and allows radio communication on a global scale. The presence of a magnetic field in the ionosphere affects radio wave propagation and the conditions of wave reflection. When the UV radiation from the Sun experiences fluctuations, or when more violent variations in the magnetic field occur, due to eruptions on the solar surface, fading phenomena or even black-outs in radio communications may occur. Such disturbances seem to be common in connection with increased Sun spot activities. They demonstrate how dependent we are on solar–terrestrial relationships. We can *hear* the effects of eruptions and other enhanced solar activity!

The maximum density of ionospheric plasmas is of the order of 10^6 particles per cm^3. The corresponding critical or cut-off frequency, i.e. the lowest frequency of waves that can penetrate the plasma, is 10^7 Hz. These figures refer to the lowest of several existing ionospheric layers, the so-called E-layer, the abbreviation for the electrically conducting layer, the others being denoted by F_1, F_2 etc, having lower plasma densities and being more sensitive to perturbations. Radio waves used for communication should therefore have lower frequencies to enable them to be reflected by the ionosphere. Signals from spacecraft entering the Earth's atmosphere might be cut off due to the high plasma density caused by increased ionization due to frictional heating of the atmosphere during re-entry.

Localized fluctuations of the ionosphere might, furthermore, introduce structured fading in radio communications. It has been found by instrumented rocket payloads (1986, 1992) and confirmed in 1994 by instruments on the Freja satellite that localized wave packets in the lower-hybrid wave frequency domain exist in the ionosphere. Local density depletions are often found to be associated with the observed wave packets. The structures can accordingly be interpreted as wave-filled cavities, which as a rule are strongly elongated along the magnetic field. The lower-hybrid frequency waves do generally depend on space-charge effects as well as on magnetic field effects. The static magnetic field introduces couplings between longitudinal plasma oscillations and transverse ion oscillations. The same type of wave has been studied in great detail in connection with fusion plasma experiments on heating and current drive. The waves have been found to be serious candidates for heating and current drive in future fusion reactors. The structure of observed lower-hybrid wave packets often closely resembles an envelope soliton, with a characteristic bump-like shape set up by the balance between nonlinear forces and dispersion. A wavelet transform technique (see later) has been used to analyse the electric field fluctuations. The wave frequency in the Freja observations is centred at 4 kHz and covers a rather narrow frequency range. The observed distributions seem to be surprisingly constant in time and vary only a couple of times over several years. The generation mechanisms of lower-hybrid waves have attracted considerable interest recently and have been summarized in the literature [4.21].

This section indicates that there are still interesting phenomena to be discovered in the ionosphere, the first space plasma to be systematically studied with regard to the propagation of electromagnetic waves in the presence of a magnetic field, studies for which Sir Edward Appleton was awarded the Nobel prize for physics in 1947.

4.2.4 Auroras

The explanation of auroras is simple, in principle. They are phenomena caused by the solar wind electrons which enter the Earth's magnetic field in the polar regions. The electrons excite the atoms in the lower part of the ionosphere from about 100 kilometres up to several hundred kilometres. When the excited atoms return to their original states they emit radiation of different colours in the spectrum. There are exceedingly high powers related to the auroral radiation. The amount of power associated with an average auroral arc is comparable with the power generated by a hydrogen bomb.

The auroras, or as they are also called the Northern Lights, have attracted the interest of generations through many centuries from mythology to the space age. Aristoteles (384–322 BC) is said to have given the following explanation of the aurora: 'The phenomena occur when air collides with fire'. Changing air for gas and fire for plasma he was in fact correct! He thought that auroras resembled 'jumping ghosts'. To describe the exceptional beauty of the auroras is, however, less simple. They should be seen! An aurora is an enormous multicoloured dynamic firework covering a large part of the sky, a three-dimensional image of the happenings occurring on the solar surface projected on our atmosphere. As modestly expressed by Chanchal Uberoi, the famous Indian expert on geocosmophysics, the auroras are like 'flying sarees in the sky'.

Present-day research is very much concerned with the questions of how the solar wind electrons which reach the aurora have been accelerated to the energies necessary to excite the atoms in the ionosphere and produce the aurora. There are indications from recent experiments by rockets and satellites that electron acceleration occurs during the passage of the electrons from the magnetosphere into the upper atmosphere, particularly in the region covering the last 10 000 km. It seems that the acceleration occurs predominantly in the direction of the magnetic field [4.22].

There are various candidates for the mechanisms which are active in accelerating the electrons, from the possibility that they have been accelerated by potential fields, to others where resonant interaction of the electrons with waves is responsible for the acceleration. The resonant interaction of an electron with a certain type of wave, the phase velocity of which may depend on the plasma density, the strength and direction of the static magnetic field etc, could be compared with the action of a surfer riding the ridge of a water wave and choosing a certain direction for his surf board in descending the whirl of the breaking wave.

A particularly interesting and promising possibility for electron acceleration by waves seems to be the interaction with lower-hybrid waves (see section 4.2.3) or bunches of such waves. The phase velocity of the lower-hybrid wave could be in resonance with the parallel motion of the electrons and simultaneously with the ions moving perpendicular to the magnetic field. It is interesting to notice how the lower-hybrid waves, which are extensively studied and used for fusion plasma heating and current drive of fusion plasmas, might play an important role in geocosmology. Electron cyclotron waves and Alfvén waves are both also important in natural and man-made plasmas.

4.3   Beyond the Sun

4.3.1   Supernovas

Heavy stars may undergo gigantic explosions and become supernovas. Though these are rare phenomena they may be the source of all the heavy elements in the universe, including those vital for life. The explosions give rise to radiation of extreme brightness. Often visible to the naked eye, the brightness of the supernova may increase at a rate of a hundred thousand times in a few hours, reaching a state which is 10 billion times brighter than our Sun. Gaseous clouds are ejected from the explosion with velocities of about 10 000 kilometres per hour. They appear in the sky as expanding nebulae in white, blue and red. A supernova explosion differs from a nova explosion in that the whole star participates, not only an outer shell.

Supernovas attracted great interest as much as 2000 years ago, when they were observed by Chinese astronomers. Annals from those days recount supernova events in the year 5 BC, and the years AD 185, 369, 1006, 1054 (Crab), 1572 (Tycho Brahe), 1604 (Kepler). The most famous of these supernovas is the one from 1054, the brightness of which was so intense that it could be seen for several months even in the middle of the day.

In the southern hemisphere a new supernova was observed in 1987. The first one in 383 years which could be seen by the naked eye, it was a comparatively short distance away—just 170 000 light-years. Observations of this supernova, SN 1987 A, have contributed to our understanding of the physics of supernovas.

It turns out that when the fuel in the chain of fusion processes is consumed the inner part of the star contracts and finally collapses.   The inner part includes elements, masses which are multiples of four up to iron, ^{56}Fe; this inner part is believed to form a neutron star. Meanwhile a shock wave with a velocity of 30000 kilometres per second expands and drives the explosion outwards. In the process of explosion the gas is supposed to be in a plasma state. It moves under the influence of the magnetic field of the supernova. The motion may create magnetohydrodynamic waves influencing the dynamics of the system. The nebulas, which remain as luminous gaseous clouds expanding out from the position of the supernova explosion, show a filamentary magnetic structure as determined from polarization measurements, for example in the Crab nebula. The magnetic fields may be interpreted as extensions of the explosion. Under certain circumstances the explosion becomes obstructed and the star develops into a black hole. But most turn into neutron stars which may continue to emit regular bursts of radiation as pulsars.

4.3.2 Pulsars: lighthouses of the cosmos

Pulsars are stars that generate electromagnetic radiation emitted in very short pulses with intervals which are extremely regular. Pulsar radiation comes, it is thought, from neutron stars which rotate with high velocity, in certain cases several hundred times per second. The direction of rotation of the neutron star makes a fixed angle with the direction of the strong magnetic field of the star. The rotation accelerates electrons in the neighbourhood, which emit electromagnetic radiation continuously over the whole spectrum from radio to visible, x-ray and even gamma ray wavelengths. Besides, the neutron star itself emits short pulses with periods corresponding to the frequency of rotation, one pulse occurring every time the beam of radiation sweeps across the Earth.

Plasma physics phenomena play a role in connection with the acceleration of electrons to almost relativistic energies. Strong magnetic fields emanate from the magnetic poles of the star and channel the radiation from the pulsar. Synchrotron radiation is produced by spiralling of the relativistic electrons in the magnetic field.

A neutron star, assumed to be formed in a supernova explosion, is indeed a very strange object. It has a mass about the same as that of the Sun concentrated into a sphere with a radius of only about 10–15 kilometres. It corresponds to a density of about 10^{14} g cm^{-1} or 100 million tons of superdense material in the volume of a teaspoon. The discovery of the first pulsar was, in fact, made by chance in Cambridge in 1967 during the course of a study of the influence of the outer corona of the Sun on the radiation from remote point sources. Subsequently, radioastronomers aimed their telescopes towards the centre of the Crab nebula, the magnificent glaring gaseous remnant of the supernova event that is known, from Chinese annals, to have occurred in AD 1054, and found a pulsar in the centre of the nebula at the expected point of the origin of the supernova explosion.

The expansion velocity of the gaseous filaments in the Crab is 1300 km s^{-1}. The time from the explosion is in accordance with the present spatial distribution of the Crab and the measured velocities. The filamentary density structure is accompanied by a magnetic field structure where the direction of the magnetic field lines follows the gas filaments, as determined by polarization measurements.

Cases where two pulsars rotate around each other at close distance have also been observed. In the process of radiation they lose energy and come slightly closer on each turn, finally ending up in collision. One might expect that the two masses of superdense material join each other by gravitational effects to form a black hole. In 1974 two American astronomers Russell Hulse and Joseph Taylor detected such a double pulsar with the radio telescope at Arecibo in Puerto Rico. Following the evolution of this rare object for several years allowed them to test the theory of relativity. They were able to verify that the double pulsars, having masses 1.40 and 1.42 times the solar mass, lost energy at a rate predicted by the general theory, or with 76 m s^{-1} annual decrease of the orbital velocity. Their investigations simultaneously proved the existence of gravitational waves. Their discoveries led to the award of the 1993 Nobel prize for physics.

The space epic *Aniara* by the great Swedish author Harry Martinson (1904–78), awarded the Nobel prize for literature in 1974, recounts the fascinating observations by the space-ship Aniara. It records and predicts the evolution of the cosmos. From Harry Martinson's poetry may we quote the following lines

World clocks tick and space gleams
everything changes place and order.

One may interpret the ticking world clocks in the gleaming space as the pulsars, rapidly rotating remnants of supernovas carrying superdense material and sending out repetitive short pulses or radiation. The discovery of the first pulsar was made much later than the poem and was recognized by the award of the Nobel prize for physics to the British radioastronomer Anthony Hewish in 1974. The dynamic structure of the universe corresponds well to the poet's observations that everything changes place and order.

4.3.3 Quasars

The universe has never ceased to astonish us. Every now and then new objects with unexpected properties are discovered. In 1960 a new type of radio source, quasars or quasi-stellar sources, were observed for the first time. At present several hundreds of such sources are known. They seem to be sources of extremely high luminosity corresponding to hundreds or thousands of galaxies, and of cosmically very small dimensions, only some light-years. One believes that the most remote quasars are at distances of the order of 14 billion light-years from us, and that they may be the central nuclei of galaxies, originating from stars and gas-clouds attracted by a black hole. They are extremely massive objects with enormous emission of energy.

Studies of a particular source 3C-273 (3C denotes the third Cambridge catalogue of radio sources) in the optical domain have revealed that the anomalously large widths of the spectral lines from quasars are connected with a very strong shift towards the red part of the spectrum. This indicates that the source is moving away from us at high speed. The same source has also been studied in the x-ray part of the spectrum with a telescope on the satellite Einstein. If the spectrum is caused by the relative motion of the quasar it moves away from us at about 80% of the velocity of light.

Provided the estimated distance of the quasar is correct (1.8 billion light-years) the source 3C-273 radiates an energy of 10^{40} joules, which corresponds to the radiation from about 1000 normal galaxies. As is sometimes the case with radio sources quasars may occur as double objects. The radiation from them, optical as well as radio, may vary in time periodically or irregularly.

It has been suggested, as an alternative, that quasars are objects at relatively close distance which have been thrown out from nuclei of exploding galaxies. Many unanswered questions still remain. So far quasars seem, however, most likely to be very remote, extremely strong sources of radiation.

The most distant quasar known so far (PC 1247+3406) has been photographed with the largest optical telescope in the world with a diameter of 9.8 m, situated at the top of the volcano Mauna Kea at an altitude of 4150 m in the Hawaiian islands. The light received from this quasar was emitted when the universe was not yet one billion years old. The photo also reveals some other similar faint spots from objects at extremely remote distances.

What can we say about the possible mechanisms of formation of those distant sources of radiation? Does plasma physics enter the formation of quasars and if so in what way? Even if astrophysical objects in general are formed by self-gravitation there are objects that cannot be explained by such processes. The reason is that the internal energy is larger than the gravitational energy. Quasars seem to belong to this category. Therefore one has to look for processes of condensation other than gravitation.

Alfvén waves may be important in the formation of quasar clouds due to a thermal instability which may occur in the presence of Alfvén waves. It is well known that magnetic fields exist in active galactic nuclei from observations of strong emissions of synchrotron radiation. The observations also show that the radiation from the nuclei is highly variable due to some strong perturbations, presumably in the magnetic field. Magnetic field perturbations create Alfvén waves. These are supposed to be damped by resonance surface damping on a thermal instability for the formation of quasar emission-line clouds. It is believed that such line-emitting clouds are small and embedded in a broad-line region, which is in pressure equilibrium with the small clouds for a temperature of 2×10^7 degrees [4.14].

The energy release in quasars is so enormous that, in fact, nuclear energy is wholly insufficient to explain what happens. The property of quasars to *eject* matter, often as jets, is an indication in favour of annihilation of matter as a source of energy. The field is open for speculation and the future may be full of surprises.

References

[4.1] Alfvén H 1942 Existence of electromagnetic–hydrodynamic waves *Nature* **150** (3805) 405

[4.2] Alfvén H 1950 *Cosmic Electrodynamics* (Oxford: Oxford University Press)

[4.3] Alfvén H and Fälthammar C-G 1963 *Cosmic Electrodynamics* (Oxford: Clarendon)

[4.4] Alfvén H 1982 Paradigm transition in cosmic plasma physics *Phys. Scr.* T **2/1** 10

[4.5] Alfvén H 1986 The plasma universe *Phys. Today* **33** (9) 22

[4.6] Lundquist S 1949 Experimental demonstration of magneto-hydrodynamic waves *Nature* **164** 145

[4.7] Lundquist S 1949 Experimental investigations of magneto-hydrodynamic waves *Phys. Rev.* **76** 1805

[4.8] Åström E 1950 Magnetohydro-dynamic waves in a plasma *Nature* **165** 1019

[4.9] Åström E 1950 On waves in an ionized gas *Ark. Fys.* **2** 443

[4.10] Lehnert B 1954 Magneto-hydrodynamic waves in liquid sodium *Phys. Rev.* **94** 815

[4.11] Chandrasekhar S and Fermi E 1953 Magnetic fields in spiral arms *Astrophys. J.* **118** 113

[4.12] Chandrasekhar S and Fermi E 1953 Problems of gravitational stability in the presence of a magnetic field *Astrophys. J.* **118** 116

[4.13] Chandrasekhar S 1961 *Hydrodynamic and Hydromagnetic Stability* (Oxford: Oxford University Press)

[4.14] de Azevedo C A *et al* (eds) 1995 Alfvén waves in cosmic and laboratory plasmas: proceedings of the international workshop on Alfvén waves (Rio de Janeiro, Brazil) *Phys. Scr.* T **60**

[4.15] Fälthammar C-G 1995 Hannes Alfvén *Phys. Scr.* T **60** 7

[4.16] Shukla P K and Stenflo L 1995 Nonlinear Alfvén waves *Phys. Scr.* T **60** 32

[4.17] Tajima T and Shibata K 1997 *Plasma Astrophysics* (Reading, MA: Addison Wesley)

[4.18] Dawson J M, Bingham R and Shapiro V D 1997 X-rays from comet Hyakutake *Plasma Phys. Control. Fusion* A **39** 185

[4.19] Levy D H 1995 *The Quest for Comets* (Oxford: Oxford University Press)

[4.20] Huebner W F (ed) 1990 *Physics and Chemistry of Comets* (New York: Springer)

[4.21] Pécseli H L, Lybekk B, Trulsen J and Eriksson A 1997 Lower-hybrid wave cavities detected by instrumented spacecrafts *Plasma Phys. Control. Fusion* **39** 1227

[4.22] Bryant D 1999 *Electron Acceleration in the Aurora and Beyond* (Bristol: Institute of Physics Publishing)

Chapter 5

Electrical discharges

5.1 Applications of plasmas

Together with auroras, lightning discharges belong to the earliest examples of plasma phenomena observed in nature. Lightning occurs in connection with thunderstorms. It often appears as a network of white, luminous inverted-tree-shaped structures or single strikes between a cloud in the atmosphere and the ground, or between two clouds. The charged particle densities in the current channel are high, about 10^{15}–10^{20} particles cm^{-3}, and temperatures of about 10^5 degrees are reached (see plate 11).

According to the National Lightning Detection Network, Tucson, Arizona:

> Hundreds of millions of dollars in property damage are attributed to lightning strikes every year in the US. Thunderstorms are the single most destructive threat to maintaining uninterrupted service to power utility customers. Lightning and electrical clouds pose a severe hazard for NASA space launches as evidenced by direct lightning strikes which damaged the Apollo 12 moon flight and destroyed an unmanned Atlas-Centaur rocket in 1987. Long-term projects exist to develop a lightning ground strike database for the entire US.

Apart from observations of atmospheric lightning, experiments are carried out in high-voltage laboratories to investigate the properties of discharges and to improve things such as isolator materials.

Lightning in the atmosphere can excite a particular type of audible frequency electromagnetic wave, so-called whistlers, which are able to be propagated over long distances in the ionospheric plasma along geomagnetic field lines, and even make several passages between the north and the south pole, producing repeated whistling signals in a radio system.

In several kinds of space device like rockets and missiles, jet plumes of plasma are formed behind the spacecraft. These can interfere with radio communications in the spacecraft and even cause black-outs in communication and signal processing. In high-voltage engineering, plasmas streaming between parallel condensor plates, and at the same time perpendicular to a magnetic field, have been used to generate electric power in so-called magnetohydrodynamic (MHD) generators. Power stations based on the MHD principle have been built, particularly in Russia.

In the field of microwave electronics electron beams have been widely used as amplifiers, using the interaction of the electron beam with a slow-wave structure, for example a helical metal wire in a travelling wave tube, a device which is used in communication satellites for example.

A type of generator of electromagnetic radiation, the free electron laser (FEL), uses the interaction of a relativistic electron beam with a periodic magnetic structure of alternating magnetic polarities, or alternatively with a microwave. The device may be considered as an application of nonlinear three-wave interaction (two transverse electromagnetic waves and one longitudinal plasma wave); high-power tunable electromagnetic radiation can be generated with free-electron lasers over a very wide frequency spectrum covering the domain from microwaves to visible light and possibly even higher frequencies. Frequency tuning is achieved by changing the accelerating voltage controlling the velocity of the relativistic electron beam. One can imagine a wide variety of applications of FELs in physics, chemistry, biology and medicine. A combination of two FELs has been used in fusion

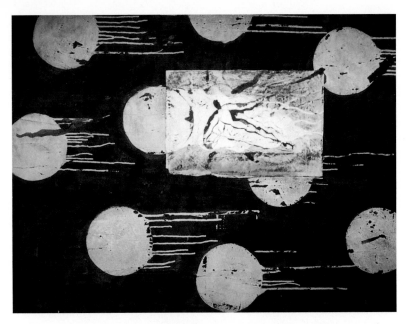

Plate 1. Cosmos: 1997 painting by Pierre-Marie Brisson. (Courtesy of Galerie Lucette Herzog, Paris.)

Plate 2. Starry Night: 1889 oil painting by Vincent Van Gogh. (Courtesy of Museum of Modern Art, New York.)

Plate 3. *Path with Cypresses and Stars: 1890 oil painting by Vincent Van Gogh. (Courtesy of Kroeller-Müller Museum, Otterlo.)*

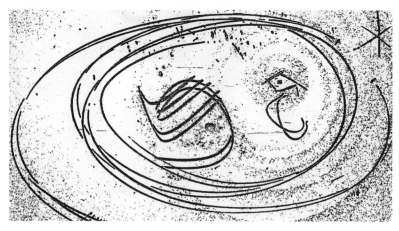

Plate 4. *Maximiliana: 1964 gravure by Max Ernst. (Courtesy of Hans Wilhelmsson.)*

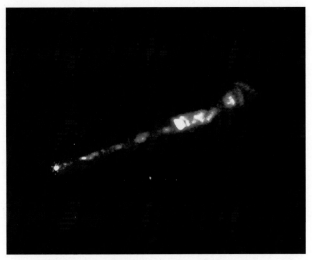

Plate 5. 'NASA's Hubble Space Telescope yields clear view of optical jet in galaxy M87'. The jet is 4000 light years long. (Courtesy of F Duccio Macchetto (ESA) and NASA.)

Plate 6. 'Hubble Space Telescope resolves braided galactic jet'. The jet is 10 000 light years long. (Courtesy of F Duccio Macchetto (ESA) and NASA.)

Plate 7. Chromolithography of a prominance on the Sun at 10 a.m., 29 April 1872 in Astronomie Populaire (1880) by Camille Flammarion. The prominance extends 200 000 km into space.

Plate 8. Chromolithography of a prominance on the Sun at 10 a.m., 15 April 1872 in Astronomie Populaire (1880) by Camille Flammarion. The prominance extends 200 000 km into space.

Plate 9. 'Photo illustration of comet P/Shoemaker-Levy 9 and planet Jupiter'. (Courtesy of H A Weaver, T E Smith (STSci), J T Trauger, R W Evans (JPL) and NASA.)

Plate 10. Comet Hale–Bopp observed on 27 March 1997. The photograph was taken over five minutes with a 200 mm tele-objective. (Courtesy of Alexis Brandeker, Stockholms Observatorium.)

Plate 11. Lightning over Bordeaux. (Courtesy of Jöel Lafon, 'Speed Photo', Bordeaux.)

Plate 12. The Wave: 1831 etching by Katsushika Hokusai.

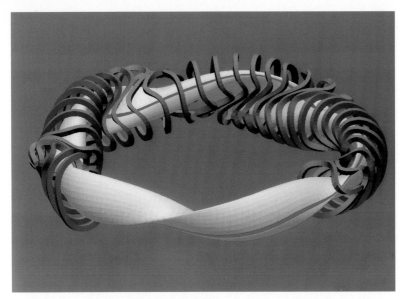

Plate 13. Schematic design of a stellerator plasma and magnetic coils. (Courtesy of Freidrich Wagner, Max-Planck-Institut für Plasmaphysik.)

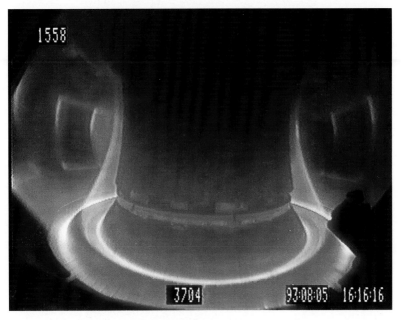

Plate 14. Plasma in the ASDEX Upgrade tokamak. (Courtesy of Freidrich Wagner, Max-Planck-Institut für Plasmaphysik.)

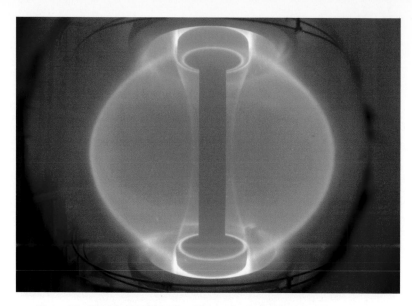

Plate 15. Plasma in START—the small, tight aspect ratio tokamak. (Courtesy of Alan Sykes, UKAEA, Culham Laboratory.)

Plate 16. Plasma in JET—the Joint European Torus. (Courtesy of Hans Lingertat and JET.)

plasma research to shape the spatial plasma current profile in a fusion device (at the Livermore National Laboratory, USA).

Plasma etching is a new and promising technique where plasmas are used in microelectronic circuit production [5.1]. Other branches where plasmas are used in industrial applications are for waste burning in plasma ovens and for surface hardening of metals. Plasma beams for cleaning stone buildings, particularly those with fine details, have come into common use. A new type of television screen, where a plasma, in fact densely packed small plasma stars, is created between two parallel glass plates, and where the image pattern is controlled by perpendicular electrodes in a quadratic system around the thin but large plasma slab, has been created recently to produce large, thin TV and computer screens, replacing conventional tubes [5.2].

References

[5.1] Reece Roth J *Industrial Plasma Engineering, Volume 2, Applications* (Bristol: Institute of Physics Publishing) to be published

[5.2] Bradu P 1999 *L'univers des Plasmas. Du Big Bang aux Technologies du IIIe Millénaire* (Paris: Flammarion)

PART 2

FUSION ON EARTH

Chapter 6

Dynamic fine structure of plasmas

6.1 Modelling as a tool for interpretation and prediction

Why is it necessary to talk so much about fusion *plasmas*? Why not simply explain how a fusion machine works? The reason is that the plasma is an integral part of a fusion machine, the part where the crucial things happen, where improvements are focussed and where benefits of operation are expected. The plasma contains the energy-producing elements, tritium and deuterium nuclei, for a reactor. It has to be confined in the interior of the vessel and suspended by 'invisible strings', preventing it from coming into contact with the walls. What happens in the bottle? How can the genie be prevented from leaking out? Since gravitational forces like those which keep the Sun together cannot be used, the plasma has to be controlled by magnetic forces.

A magnetic bottle having a straight cylindrical geometry will leak at the ends. Therefore why not bend the cylinder to give it the form of a torus like the shape of a car tyre or a ring doughnut. The charged particles will move in a curved magnetic field structure, which is also inhomogeneous in space. The density of the charged particles, as well as the current density which they produce when they move in the toroidal magnetic field, will also be inhomogeneous in space, as will the temperature of the plasma. All these quantities will have more or less bell-shaped structures in the cross section of the torus. The way in which the plasma forms certain preferred shapes and tends to remain in those shapes has sometimes been referred to as profile consistency. This tendency may be considered as an example of the means by which plasmas exploit their inherent complex features to produce particular forms or structures in the presence of the confining magnetic field. The parameters which account for particle and heat transport in the fusion plasma, i.e. the particle diffusion and the heat conductivity parameters, depend on phenomena which may be called by the common term *dynamic fine structure of the plasma.*

The analysis of such phenomena requires sophisticated mathematical techniques and computer simulations. To connect the results of analysis with experimental results in the laboratory one has to resort to certain strategies. The connection can be twofold, either an *interpretation* of observed experimental results by means of analyses, or a *prediction* of new results to guide new and more advanced experiments. The aim is always to obtain a better understanding of the processes and better performance of the reactor. The predictions could be based on new and extended analysis and experiments, on the extrapolation of previous experiments or on a combination of the two. Such procedures are highly relevant to the design of the next generation of large fusion experiments. Advances in both interpretation and prediction are related to the concept of modelling as a tool to visualize what happens or what is going to happen in an experiment.

Modelling often refers to the use of mechanical models, since we are used to such models from real life. In this respect modelling may transfer certain complicated or abstract scenarios to situations which are more easily understood due to experience. Modelling could also refer to selecting important elements of a complex system in order to describe the essential features of its behaviour.

The modelling of fine structure phenomena and their role in the global behaviour of the whole plasma is basic to fusion plasma science and technology. In this connection modelling often consists of describing a phenomenon or a scenario in mathematical terms, i.e. in formulating an equation or equations in terms of variables which correspond to observable quantities, such as densities, temperatures and magnetic fields, which may all change in space and time. The next step would be to solve the equations by mathematical analysis or by computer simulation. Simplified analysis can often be useful to test even extremely complicated computer programs at check-points.

6.2   Waves and instabilities

Why is it that the field of waves and instabilities has constantly attracted such great interest since the very beginning of fusion plasma research? One reason is that these phenomena may have very important consequences for the behaviour of fusion plasmas, and another is that they offer many challenging basic scientific problems.

A fusion plasma can carry a large variety of different types of wave, such as plasma waves, electromagnetic waves, ion-cyclotron waves, electron-cyclotron waves, hybrid waves, magnetoacoustic waves, magnetohydrodynamic waves (Alfvén waves) etc [6.1]. From a practical point of view certain waves are particularly important, because heating of the plasma to high temperatures by external sources can be achieved by exciting the plasma at certain frequencies which are characteristic of those particular waves. As examples low-frequency ion-cyclotron waves or high-frequency electron-cyclotron waves can be used for this purpose, the excitation frequencies being determined by the magnetic field and the masses of the particles. For hybrid waves the frequency depends on a combination of the ion and electron masses. Such waves are also candidates for practical heating of the plasma, as are Alfvén waves, low-frequency waves whose properties depend on the magnetic field and the density of the plasma. In fact, the properties of Alfvén waves can be deduced by proper modelling of this type of wave in terms of a mechanical analogy, namely a vibrating string. The mass and tension of the string, in this analogy, would play the role of the density and the magnetic field in the plasma.

Such low-frequency waves are variously called magnetohydrodynamic (MHD), hydromagnetic or Alfvén waves. They have many interesting properties; a basic one is that in the associated motion of the charged particles, the perpendicular velocity component with regard to the magnetostatic field is approximately the same for both ions and electrons, in spite of their very different masses. As a consequence, the gas moves essentially as a whole across the magnetic field.

The essence of magnetohydrodynamic phenomena can be traced back to the following situation: if an electrically conductive gas (a plasma) or liquid moves in a magnetic field an interaction between the magnetic field and the electric currents occurs which affects the motions of the gas or the liquid and also the magnetic field. Accordingly a coupling becomes established between the magnetic and hydrodynamic phenomena—thus the name magnetohydrodynamic phenomena. The Swedish scientist Hannes Alfvén received the Nobel prize for physics in 1970 for his work in the field of magnetohydrodynamics. These phenomena are of vast astrophysical interest and have important consequences for laboratory plasmas.

Certain phenomena like Sun spots, which can be seen occasionally by the naked eye as dark regions on the solar surface, may be explained, at least in part, by the assumption that Alfvén waves are created by perturbations in the central parts of the Sun and propagate to the solar surface where the spots are formed.

Other characteristic types of wave, plasma waves, which may be regarded as electroelastic types of wave, are fundamental in plasmas and are of interest also in plasma diagnostics. These waves are longitudinal in nature, i.e. the particle oscillations occur in the direction of the wave motion.

The way in which electromagnetic waves propagate in plasmas has important consequences for communication technology. Radio waves are scattered by the ionosphere, the electron plasma surrounding the Earth, and propagate around the Earth as a result of this scattering. Electromagnetic waves, microwaves of centimetre to millimetre wavelengths, can be used to probe fusion plasmas to determine their spatial density distribution. Such waves are transverse waves, i.e. the particle oscillations are transverse (or perpendicular) to the direction of propagation of the waves. When propagating in a plasma they are affected by the presence of a magnetic field. If the plasma density in a certain region is high enough, i.e. if the plasma frequency v_p is greater than the frequency of the wave, then the wave cannot propagate further but becomes reflected. Phenomena of this type cause interruptions to radio communications, for example when spacecraft enter and ionize the Earth's atmosphere.

Among Nature's elements none caught the interest of Leonardo da Vinci more than water. His note-books are full of sketches of waves. He studied how waves form, how they turn and whirl and create bubbles. With the same precision he drew or painted waves in human hair as in water on the sea. The waves that he studied on the shore he connected with the motion of the wings of a bird or breathing in a person. He devoted much attention to waves entering shallow water and being reflected from the shore (see figure 6.1). He made sketches of flowers and of terrifying scenes of cataclysms, i.e. the end of the world, in the same devoted way. His cataclysms resemble many active solar spots. He constructed artificial wings after watching the motions of flying birds, and he designed projectiles which were aerodynamically shaped and had directional wings which made them seem like modern rockets or precursors of the space age (figure 6.2).

Instabilities of waves or other perturbations may occur as self-amplifying phenomena in plasmas. They have been investigated extensively with regard to various plasma configurations and conditions and with regard to different kinds of wave oscillation. Such instabilities show a tendency to develop when the plasma possesses some kind of free energy, which it would prefer to get rid of and transfer to wave motion. The origin of such free energy could be an inhomogeneity in the plasma density or temperature or magnetic field. Such inhomogeneities almost always exist. The free energy could also be associated with a beam of electrons penetrating a plasma.

Figure 6.1. 'Waves entering shallow water'. Drawing by Leonardo da Vinci (Codex Madrid II 1503).

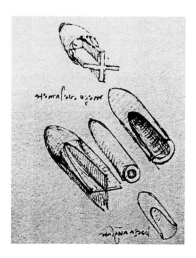

Figure 6.2. 'Aerodynamically shaped projectiles with directional wings'. Drawing by Leonardo da Vinci (Codex Arundel 1502).

Instabilities may affect the plasma configuration negatively and even spoil the confinement completely. An example of an instability which could have disastrous effects is one that could occur in a tokamak torus if the current in a plasma were to be driven by magnetic induction to such high values that the magnetic poloidal field became too strong for the plasma torus to be kept together causing it to break into wiggles. In limiting the plasma current to avoid the risk of such instabilities one also limits the heating due to plasma resistance, ohmic heating.

Nonlinear effects are important for waves and instabilities. Nonlinearities may saturate instabilities but they can also cause or enhance instabilities. We discuss these in the next section.

6.3   Nonlinear effects

For a long time nonlinear effects were considered to be small corrections to the linear response of a medium to some kind of influence. It was considered natural that when an electric voltage was applied to a wire it would cause a current proportional to the voltage. Or when an electromagnetic wave launched into a medium, for example a plasma, the wave penetrating the medium, and also the reflected wave, would have amplitudes proportional to the amplitude of the incident wave. The waves should also have the same frequency. The response was always proportional to the primary source, i.e. to the voltage or the amplitude of the incident wave, as long as the source was weak. For strong sources, i.e. above a certain threshold power, things change and new interesting phenomena may occur. The medium starts to react nonlinearly [6.2–6.5]. The nonlinear response can then become larger than the linear one, and dominate the situation. Nonlinear phenomena are becoming of growing interest in science. Historically, nonlinearities were first noticed in the 1920s in the field of chemistry. At present they are studied extensively in connection with biological problems and in several fields of physics, notably laser and plasma physics. From a mathematical point of view important developments have occurred, starting from the early work of Volterra who studied the oscillations of nonlinearly coupled systems, namely periodic changes in the abundance of fishes of different kinds in the Adriatic sea!

Milestones in nonlinear theory are the contributions of Fermi, Ulam and Pasta in 1955 on thermalization of nonlinear vibrations in a chain of periodically situated particles between which linear and quadratic forces operate, and related work by Zabusky and

Kruskal in 1965 on solitons, i.e. spatially bell-shaped distributions of exceptional stability governed by the mutual influence of two phenomena: *nonlinearity* and *dispersion*. During the last decades considerable efforts have been devoted to studies of self-organization in non-equilibrium systems and dynamics of synergetic systems, and to the problem of chaos, for which modern computer simulation has become an essential tool.

In Japanese art Hokusai (1760–1849) was one of the great masters. 'The Wave', one of his etchings in the suite entitled *Thirty-Six Views of Mount Fuji*, is a grandiose demonstration of wave-breaking, i.e. the nonlinear effect where the amplitude of a wave becomes so high that the top of the wave splits into separate filamentations and bubbles (plate 12). The impression of the force of the wave is enhanced by the design, in which the wave, seen in the foreground, takes up an area of the design which dominates Mount Fuji seen at a distance. I was told by Japanese scientists that in his lifetime Hosukai lived in about a hundred different places in Japan, that he gave away everything he could spare and that he was known as a very happy man. One cannot help comparing him with creative people of our time!

In the early 1960s the principle of light amplification by stimulated emission of radiation, the laser, was discovered. It meant that the high power of coherent light could be concentrated in small domains in space and also in short intervals of time. Excellent conditions for research on nonlinear effects were suddenly provided. The field of nonlinear research really exploded with the advent of the laser. The need for basic understanding of nonlinear phenomena became particularly pronounced when high-power laser projects started in the US and USSR, and also in Japan, accompanied by research on laser fusion, i.e. attempts to produce fusion reactions by microballoon explosions caused by laser pulses. Nonlinear effects were particularly important for the coupling of the laser pulse with the microballoon (pellet) and for making diagnostics of the pellet plasma. For the diagnostics of fine structure phenomena and chaotic behaviour of plasmas the use of ultrashort pulse techniques seems to be an interesting future possibility.

Nonlinear effects in plasmas have been studied intensively since the early 1960s when large magnetic fusion experiments also started to be constructed. The field of reaction–diffusion problems, i.e. studies of situations where reaction and diffusion processes occur simultaneously, and which are therefore governed by two phenomena, nonlinearity and diffusion (which is generally also nonlinear), has recently attracted considerable attention. Numerous applications of reaction–diffusion phenomena are found in modern science, for example in fusion plasma physics where the evolution of a burning fusion plasma is governed essentially by nonlinear diffusion and by alpha particle heating. The problems of heat and particle transport and of magnetic confinement in toroidal devices are key questions for the realization of a future fusion reactor [6.6–6.13].

6.4   Three-wave interaction

If three waves are present simultaneously in a medium, all of them sufficiently intense that each one experiences the presence of the two others, interesting things may happen. One obtains a three-wave interaction.

From a nonlinear point of view the phenomenon is a particularly interesting one for the following reasons: firstly, it is complex enough to exhibit features which it may have in common with more general systems, where many waves interact, like a plasma turbulence, and secondly it is sufficiently simple from an analytical point of view that it can be solved exactly [6.5]. There are many different types of wave which may occur in plasmas, some closely related to the motion of the electrons, some to the motion of the much heavier ions, some determined by the presence of a magnetic field, like ion- or electron-cyclotron waves; others may be electromagnetic waves. Therefore it might be possible to excite a wide spectrum of waves from low frequencies to high frequencies, all fulfilling an equation of the medium relating the wavelength to the frequency of the medium, the so-called dispersion relation. When two waves interact to form a new wave not only the primary waves but also the produced wave should be natural waves of the medium. This means strong selection of the possible ways of coupling. Only the right selection gives what is called *resonant* interaction with possible strong excitation of the new wave. It should be remembered here that we are discussing nonlinear interaction of the simplest type, where the excitation of each wave depends on the product of the two other waves.

It is easy to imagine a simple mechanical analogy to resonant coupling. If a plate or membrane is exposed to two sources of vibration of different frequencies a mutual coupling of these perturbations may occur. Only if the sum (or difference) frequency coincides with another natural frequency of the plate may the coupling have a noticeable effect as a result of resonant interaction. The same may apply to the rotating tyres of a car tested by holography!

When lasers became available in the early 1960s beautiful experiments were carried out on resonant three-wave interaction. An experiment was done in a birefringent crystal which demonstrated that resonant coupling could occur only on the surface of a cone, where the matching conditions for the frequencies and wave numbers were fulfilled. The result was a beautiful luminescent conical surface in agreement with theory. Similar experiments have been performed in magnetized plasmas in the microwave domain of the spectrum.

It turns out that three-wave interaction can be an unstable process, a simple example of nonlinear instability. A necessary condition for this to happen can be shown to be that the medium, for example a plasma, possesses free energy. This is the case when a plasma is penetrated by an electron beam having kinetic energy. Such a combined plasma–electron beam medium also has the interesting property that one of the three interacting waves carries negative wave energy. This means that the total wave energy (mechanical plus electromagnetic) counted per unit volume of the medium is lower in the presence of the wave than in the absence of the wave. The concept of negative wave energy has no practical sense, except when other waves are present simultaneously, which carry positive wave energy. Since this is, in fact, the case for the other two waves in the three-wave interaction considered, a most extraordinary phenomenon may occur, namely that all three waves grow in amplitude simultaneously and conserve the total wave energy of the three waves in the process of growth. It turns out that the process will be an explosive three-wave interaction. All the three waves will grow explosively together to approach exceedingly high values in a finite time.

A mechanical, or rather acoustical, analogue, may illustrate the unstable three-wave process. Suppose that two people are getting into serious conversation and their voices are increasing in amplitude. A third person who has taken interest in the conversation wants to join the discussion. To be noticed he has to speak up loudly. This causes the others to raise their voices even more, each of them defending their opinions *vis-à-vis* the others, and so on, the state reaching a possibly violent collapse in a finite time!

Other cases of three-wave interaction may not be of the explosive type. However, it has been demonstrated that such cases may lead to other interesting situations. Imagine that one wave has linear growth (i.e. for small amplitudes), and that the two other waves are linearly damped. It turns out that if the linear growth is considerably higher than the damping *chaotic* behaviour of the system is caused as a result of the competition between the nonlinearly coupled waves, and the solution breaks into numerous new branches.

The examples given here illustrate how simple nonlinear processes may have rather remarkable effects.

The dynamics of fusion plasma is governed by a set of competing processes. The change in temperature of the plasma depends on the heating from the fusion reactions and from the other auxiliary heating processes; it depends on the heat conduction, and losses, all these processes being of nonlinear nature. They may help to set up a stable state by balancing each other, but they could also hinder the establishment of such a state were it not for the presence of the confining magnetic field and for the influence of the boundaries.

In fact, even the simplest forms of reaction–diffusion equations show tendencies to self-formation, corresponding to self-organization in more complex structures, and of pattern formation as in biological and chemical systems.   Profile consistency may be regarded as a consequence of such tendencies in a fusion plasma.

6.5   Evolution of populations: explosive instabilities

The problem of evolution of populations plays an important role in physics, chemistry, biology and several other fields of science. Studies of related questions generally lead to systems of coupled nonlinear differential equations, which are most often handled by computer analysis.

A simple example could be mentioned here, which in spite of its simplicity has important consequences. It expresses how the change of a dynamic variable, let us say u, for example the density or temperature of a medium, per unit time is proportional to u squared which results in an interesting dependence of the variable on time, namely that the variable increases steadily towards very high, in fact infinite, values within a finite time [6.14].   The time dependence, therefore, has an explosive trend. (It might be an interesting exercise to prove that this singular behaviour is described by $(t_0 - t)^{-1}$ where t_0 is the time of explosion.)

The time it takes for the variable (the temperature or density) to reach infinitely high values is shorter the larger the initial value and the stronger the nonlinear coupling. (Again readers may like to prove this statement for themselves.)

It has been suggested that the observed explosive increase in field energies and temperature of solar flares (pulses of hard x-rays, prompt gamma-ray lines and microwaves, simultaneous and similar in shape) is caused by current loop coalescence of two magnetic islands. As a result the magnetic field energy, which is proportional to the strength of the field squared, the electrostatic energy, proportional to the electric field strength squared and the temperature, T, all diverge explosively as a function of time (see [4.17]).

Since the quadratic term u^2 may be interpreted as a natural source term for the growth of population n (compare a sink term such as $n_e n_i$ for the recombination of electrons and ions in a plasma, or n_e^2 for $n_i = n_e$) it seems tempting to make some comparisons with empirical data, and possibly some predictions for the future.

Empirical data over several hundred years show that the increase in human population fits closely to an evolution of the explosive type. In fact the drastic increase in our population during the last decades has followed astonishingly well the predictions of a simple equation with a quadratic term, accounting for the simultaneous death-rate separately by a linear term [6.15].

The prediction of the theory would be that if there is no drastic change in the trend of evolution extremely high values will soon be expected in the world population. According to United Nations estimates, the world population, which is today 5.4 billion, might reach 14 billion in a few decades.

The asymptotic infinity is estimated to occur in the third decade of the 21st century. The only efficient way of preventing an explosion of population seems to be to control the source term, thus making the parameter a strongly decreasing function of time. A reason for discussing the population problem here at some length is that it illustrates, for a simple nonlinear case, the principle of modelling. Besides, the population problem is of vital interest for our future; the nature of population increase will strongly influence the required energy per capita, and is thereby linked to the need to realize fusion energy.

Furthermore, if the variable u in the simple equation is chosen to represent the temperature T in a fusion plasma, the term T^2 could be a realistic source term for representing alpha particle heating in a fusion plasma, i.e. accounting for the contribution of helium-4 to plasma heating in fusion reactions. A full equation for the temperature evolution of a fusion plasma should also include terms accounting for the conduction of temperature and losses by radiation as well as for the influence of the plasma boundaries. Even so a simple nonlinear model could represent the influence of heating of the plasma by fusion energy. It might obviously lead to a strong evolution of temperature in time, and would crudely represent a model of the hydrogen bomb.

A fusion reactor should make efficient use of the heating process, but should also effectively confine the fusion plasma. Intensive research on the principle of magnetic confinement using experiment, analysis and computation has resulted in clear evidence that a solution to this problem is available. Today one also has reason to be confident of the future success of efforts to reach the goal on a practical level.

A question which parallels the fusion energy problem is then: will we be able to master the problem of controlling the evolution of the world's population, which seems to be the most urgent question of all? With a comparatively short time scale? Should this problem be addressed urgently to avoid an 'explosive' catastrophe?

6.6 Vortices

When cream is stirred into a cup of coffee, when a propeller moves in the sea or when a rocket takes off in the air, vortices are formed. They are common in the atmosphere and occur as galaxies of stars in the sky. They appear everywhere in nature and in technology, and also in fusion plasmas. Extensive studies of vortices have been made in fluid dynamics by experiment and analysis using a set of nonlinear partial differential equations, named Navier–Stokes equations, mostly for two-dimensional cases [6.16, 6.17].

Vortices played an important role in the early history of science. Cosmological vortices were thought to produce order out of the chaos. Democritus (500 BC) who also introduced the atom, made vortex motion a tool to formulate general physical laws. The concepts of the atom and the vortex motion together formed a primitive first 'unified theory'. Today the early ideas about vortices could be given a somewhat extended interpretation, namely that small fluctuating disturbances in the very beginning of the universe may be regarded as seed vortices from which galactic systems later evolved.

There are many shapes that vortices can take, from elongated tornadoes to flat hurricanes, more commonly observable as bath-tub vortices as bath-water goes down the plug-hole. An interesting property of vortices is their ability to produce locally high rates of rotation. In a tornado or a bath-tub vortex a fluid element has a constant angular momentum rv, where r is the radial distance of the element from the centre of the vortex and v is the circumferential velocity. When the element is drawn to the centre of the tub the velocity increases as $1/r$ which explains the violent central rotation, also accompanied by a minimum of the pressure in the centre.

Leonardo da Vinci (1452–1519) took an active interest in making detailed designs of the phenomena which occurred when a stream of water fell into a basin filled with water or when an obstacle like a flat plate was inserted into the stream. Patterns of vortices and bubbles in Leonardo's designs are reproduced today in numerical studies using large computers, which produce results which show detailed similarities to his original drawings. His studies were of practical interest in his own time in relation to a hydraulic project in Milan in 1507 (see figures 6.3 and 6.4).

Leonardo da Vinci's scientific drawings of water in motion remain today as artistic works. He studied a water jet running out of a square-shaped hole into a pond, where it caused a pattern of bubbles and whirls resembling the structure of a chrysanthemum flower. The drawing was probably made around 1507 in connection with the above-mentioned hydraulic project in Milan. The water jet may be compared to a beam of neutral particles injected into a plasma where it causes vortices and turbulence when heating the plasma (see figure 6.3).

Today we study such phenomena in great detail using computer simulations and the results are similar. Compare plasma streams around obstacles such as plates and wires in

Figure 6.3. 'Bubbles and whirls caused by a water jet falling into a pond'. Drawing by Leonardo da Vinci, probably around 1507 (Windsor, Royal Library).

laboratory experiments, or the bow-shock and the tail of the magnetosphere caused by interruption of solar wind flow around the Earth. Detailed analytical studies of vortices are simplified by the introduction of vorticity, an abstract notion in terms of which the Navier–Stokes equations can be transformed into a form which is more lucid both physically and mathematically. In terms of vorticity it can be shown that microscopic vorticity at a point in space can generate macroscopic patterns, i.e. very small-scale phenomena can develop into large-scale structures, as has been demonstrated in computer-generated images in the two-dimensional case. The origin of biological evolution can be traced back to the development of patterns due to random disturbances of the genetic code.

A generalization of the dynamics of vortices to a three-dimensional description of reality, which both in principle and computationally is a formidable task, may reveal deep secrets and produce new surprises. An indication of such

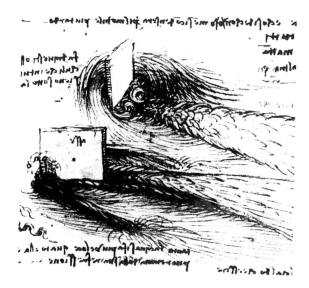

Figure 6.4. 'Whirls and wakes caused by a rigid plate introduced in flowing water'. Drawing by Leonardo da Vinci, probably around 1507 (Windsor, Royal Library).

expectations comes from daily weather forecasts which can easily fail to predict even certain large-scale phenomena which may originate from small local perturbations in an atmospherically sensitive, e.g. very inhomogeneous, domain in another part of the world.

How can these ideas be connected to fusion plasma experiments? Obviously, it would be disastrous if small perturbations were to develop into gigantic plasma motions and spoil the global confinement. But on the other hand, it could sometimes be useful for heating purposes if perturbations in certain domains of the plasma were to spread out over the whole plasma and not stay localized. This may be the case when auxiliary heating from high-frequency waves or neutral beams is applied to heat the fusion plasma.

6.7   Wavelets and turbulence

Wavelet theory is a new and interesting approach which may help improve our understanding of turbulence. Wavelets can be considered simply as mathematical tools which are generalizations of Fourier integrals. They may be used to describe the propagation and nonlinear interaction of pulses in space-time, allowing for special frequency control in the system of waves. In music, for example, such frequency control is necessary to screen the dependence in time of different parts of a composition ('windowing' by means of the pedal control of a piano). In physics, one may think that different nonlinear mechanisms or changes introduced by geometrical effects in, for example, a fluid stream can introduce effects of a similar nature. A crucial question concerning turbulent flows is how to separate coherent structures (vortices) from the rest of the flow. So far the structures responsible for the chaotic and accordingly unpredictable behaviour of turbulent flows have not yet been identified. There are strong indications from laboratory and numerical experiments that vortices are such elements. They might form the basis for constructing a new statistical mechanics in terms of equations appropriate for fully developed turbulent flows. It seems to be necessary to develop new approaches in this direction since available theory based on incompressible Navier–Stokes equations from fluid mechanics is not adequate for turbulence at high Reynolds numbers (the Reynolds number represents the ratio of the nonlinear convective motions, responsible for the flow instability, to the linear dissipative damping which converts kinetic energy into thermal energy), which cover many orders of magnitude, representing various fields from naval engineering and aeronautics to meteorology and astrophysics [6.18–6.20].

All classical methods in turbulence are based on Fourier representation which is inappropriate for a nonlinear convection term. Wavelet theory has recently been applied to analyse, for example, electric-field fluctuations of observed lower-hybrid wave packets in the ionosphere. The wavelet technique has established itself as a useful tool for studies of wave packets in various connections and should be of considerable interest in application to fusion plasmas.

6.8   From fine structure to global dynamics of fusion plasmas

When an electron vibrates the Universe trembles

Sir Arthur Eddington

How is it possible that fine structure dynamics (like the motion of bubbles in boiling water) can influence the global dynamics of fusion plasmas? Do similar situations exist elsewhere in Nature? Well, it is known that dolphins make use of small vibrations in their skin to help propel their heavy bodies through water. They even leap out of the water! The small vibrations in their skin lower the resistance of the overall motion of their bodies through the water appreciably, by as much as ten times, probably because they counterbalance the influence of water microvortices set up at the interphase between their skin and water.

A similar example may be taken from naval technology. The resistance from the water, and thereby the power used to propel the ship, can be reduced considerably by shaping the underwater prow-profile in a particular way. A correctly designed 'bump' immediately below the water-line in the prow-profile will cause a wave diffraction pattern compensating for the effects of the 'ordinary' bow-shock waves (compare this with Cherenkov radiation from fast particles in media with a velocity v greater than the speed of light c in the medium) on the surface of the wave. This artificially introduced 'skin' effect lowers the water resistance and thus the power necessary to propel the ship.

In a similar vein financial disturbances in one country can grew quickly and may even shake the global economy. A small fire can ignite explosively in the wind and may burn large forests (compare the effects of lightning). A political conflict can lead to violent agitation and may even be the origin of a world war.

From ecology it is well known that small disturbances, famously from the wings of a butterfly, in the air localized under certain conditions such as sharp temperature or density gradients, in one part of the world, could induce giant phenomena like cyclones at remote distances in the atmosphere in other places.

From these examples we notice that wave phenomena which may seem tiny could have a strong influence on the overall motion of large systems. With this in mind let us now see how small-scale phenomena may influence the large-scale motions of fusion plasmas.

It is known from laboratory experiments with fusion plasmas that plasma phenomena and the plasma structure near the plasma edge can have a decisive effect on the properties of the whole plasma, for example with relation to plasma confinement. Of the two characteristic modes of a fusion plasma, the L mode and the H mode, the H mode corresponds to steeper gradients near the edge and also provides a confinement time which could be twice that of an L mode.

To study the bulk behaviour of fusion plasmas one has to consider the influence of the various processes that contribute to the dynamics of the whole plasma. For the analysis of this problem one has to construct an equation in which the separate terms model the various processes which contribute actively to plasma behaviour. In doing so, for example in describing the temperature evolution in a plasma, one has to model the contributions from the process of temperature conduction as well as from creation processes, which have a tendency to increase the temperature by ohmic, alpha particle or external sources of heating, like high-frequency or neutral particle heating. Loss processes, for example by bremsstrahlung, also have to be taken into account. Each of these processes has a particular nonlinear dependence on temperature and density of the plasma: for example the alpha particle heating nT^p, where p is a parameter of the order of two for a hot plasma and n and T denote the density and temperature of the plasma. The temperature conduction process entering the equation is an even more complicated term involving spatial derivatives. In a one-dimensional model this term could take many forms, depending on what type of phenomenon one assumes to be responsible for temperature conduction. A particular phenomenon which is often considered is the temperature gradient-induced drift-wave instability, which causes turbulence in the plasma and leads to a temperature dependence of χ which is approximately equal to temperature (T) to the power δ where $\delta = 3/2$, a characteristic value for this type of turbulent fine structure. Losses may be represented by a bremsstrahlung term n times T to the power q where $q = 1/2$.

The equation describing the global dynamics of the plasma temperature will thus include a number of terms which depend differently on the temperature and which are all nonlinear. The sum of terms representing the individual physical processes will be equal to the rate of change of the temperature $\partial T/\partial t$, where t is time. A similar equation in n, the plasma density, can be constructed for the particle diffusion. The global dynamics of the fusion plasma will accordingly be governed by a coupled system of nonlinear partial differential equations in T and n, which may be studied by analytical or numerical simulation methods, including the possibility of phase-plane description [6.7–6.9].

The results will give information on the possibilities of equilibria, and on the details of the approach of the plasma to such equilibria as well as of the stability properties [6.21].

In fact, the partial differential equations describing the simultaneous influence of reactions, for example alpha particle heating, and diffusion processes (and therefore often called reaction–diffusion equations) also have solutions with timescales characteristic not only of diffusion but also of propagation phenomena. These can be much faster than ordinary diffusion processes. The propagating solutions depend on the spatial gradient (steepness) of the pulse or the wave, which the solutions represent. In this context it should be mentioned that studies of certain perturbations of the plasma, which give almost immediate responses at a large distance from the source of perturbation have until now lacked an explanation [6.22].

It seems that the nonlinear plasma phenomena continuously offer new and unexpected results. A satisfactory understanding of transport processes in fusion plasmas requires detailed knowledge of the complex plasma system.

References

[6.1] Stix T H 1992 *Waves in Plasmas* (New York: American Institute of Physics)

[6.2] Sagdeev R Z and Galeev A A 1969 *Nonlinear Plasma Theory* (Amsterdam: Benjamin)

[6.3] Tsytovich V N 1970 *Nonlinear Effects in Plasmas* (New York: Plenum)

[6.4] Davidson R C 1972 *Methods in Nonlinear Plasma Theory* (Amsterdam: Benjamin)

[6.5] Weiland J and Wilhelmsson H 1977 *Coherent Nonlinear Interaction of Waves in Plasmas* (Oxford: Pergamon)

[6.6] Smoller J 1983 *Shock Waves and Reaction–Diffusion Equations* (New York: Springer)

[6.7] Wilhelmsson H and Le Roux M-N 1993 Self-consistent treatment of transport in tokamak plasmas *Phys. Scr.* **48** 735

[6.8] Le Roux M-N, Weiland J and Wilhelmsson H 1992 Simulation of a coupled dynamic system of temperature and density in a fusion plasma *Phys. Scr.* **46** 457

[6.9] Wilhelmsson H, Etlicher B, Cairns R A and Leroux M-N 1992 Evolution of temperature profiles in a fusion reactor plasma *Phys. Scr.* **45** 184

[6.10] Wesson J 1987 *Tokamaks* (Oxford: Oxford University Press)

[6.11] Krall N A and Trivelpiece A W 1973 *Principles of Plasma Physics* (New York: McGraw-Hill)

[6.12] Itoh K, Itoh S-I and Fukuyama A 1999 *Transport and Structural Formation in Plasmas* (Bristol: Institute of Physics Publishing)

[6.13] Weiland J 1999 *Collective Modes in Inhomogeneous Plasmas and Advanced Fluid Theory* (Bristol: Institute of Physics Publishing)

[6.14] Wilhelmsson H 1972 On the analysis of the population problem *Phys. Scr.* **5** 116

[6.15] Weiland J and Wilhelmsson H 1974 On the evolution and saturation of the world population *Phys. Scr.* **10** 257

[6.16] Lugt H J 1985 Vortices and vorticity in fluid dynamics *Am. Sci.* **73** 162

[6.17] Lundin R and Marklund G 1995 Plasma vortex structures and the evolution of the solar system—the legacy of Hannes Alfvén *Phys. Scr.* T **60** 198

[6.18] Farge M 1992 Wavelet transforms and their applications to turbulence *Annu. Rev. Fluid Mech.* **24** 395

[6.19] Farge M, Kevlahan N, Perrier V and Goirand E 1996 Wavelets and turbulence *Proc. IEEE* **84** 639

[6.20] Lesieur M 1997 *Turbulence in Fluids* (Dordrecht: Kluwer)

[6.21] Wilhelmsson H, Lazzaro E and Cirant S 1996 Sensitivity of fusion plasma temperature profiles to localized and distributed heat sources *Phys. Scr.* **54** 385

[6.22] Lazzaro E and Wilhelmsson H 1998 Fast heat pulse propagation in hot plasmas *Phys. Plasmas* **8** 2830

Chapter 7

The art of magnetic confinement

7.1 The principle of magnetic confinement

A remarkable property of magnetic fields may be the key to the problem of fusion plasma confinement, namely the fact that free motion of charged particles is not allowed in directions transverse to the magnetic field. Instead, the particles will spiral around the magnetic field lines. The radius of the spiral, the Larmor radius, is proportional to the square root of the thermal energy and inversely proportional to the magnetic field. For typical tokamak ion temperatures $T = 10$ keV and magnetic fields $B = 5$ T the ion Larmor radius is less than a centimetre.

It turns out, furthermore, that the sum of the plasma pressure and the magnetic field pressure is a constant, equal to the magnetic field pressure outside the plasma. In the plasma the ratio of the plasma pressure and the magnetic field pressure, usually denoted by β, is less than unity and in tokamaks is typically only about 5%. Even if elementary considerations clearly indicate that magnetic fields offer attractive possibilities for the confinement of fusion plasmas important problems remain concerning the thermal flux and the influence of boundaries. Related problems are actively being studied both experimentally and theoretically.

The magnetic fields used for the confinement of the fusion plasma could be generated from outside the plasma, using a convenient set of coils, or could be produced by the plasma itself, using the magnetic field originating from the plasma current. In a tokamak, use is made of internally as well as externally generated magnetic fields, whereas in stellarators, which have no toroidal current, only externally created magnetic fields are used. As a result, tokamaks and stellarators exhibit vastly different magnetic configuration architectures, and also have alternative schemes of operation [7.1, 7.2].

Before discussing related topics further, let us emphasize some of the basic questions relating to fusion energy generation and to the conditions for self-sustained fusion.

7.2 Fusion energy generation and self-sustained fusion

The reaction between deuterium (D) and tritium (T) can be illustrated as

where $\oplus$ denotes a proton and $\bigcirc$ a neutron, or

$$D + T \rightarrow {}^5He \rightarrow {}^4He + n + 17.6\,\text{MeV}.$$

As an intermediate state a compound nucleus of ^{5}He is formed which, however, immediately splits into an alpha particle and a fast neutron, sharing the liberated energy as 3.5 MeV and 14.1 MeV, respectively. The majority of the alpha particles stay in the plasma and contribute to the plasma heating, whereas the neutrons, n, leave the plasma with enormous energy, available for useful energy production.

The D–T reaction has the highest power production rate of the fusion reactions. For T about 10 keV and $n \sim 10^{14}$ cm^{-3} and a D–T fusion cross section of 10^{-26} cm, the generated power density is about 1 W cm^{-3}.

The fusion reaction rate has to be higher than the energy losses from the plasma. Denoting the ratio between the thermal energy in the plasma and the energy losses by τ_E, the energy confinement time, one can conclude that a condition for self-sustaining fusion by D–T reactions (the Lawson criterion) is

$$n\tau_E > 2 \times 10^{14} \text{ (cm}^{-3}\text{ s)}.$$

The critical limit depends on temperature and has a maximum between 10 keV and 20 keV. The high temperature is necessary to surmount the Coulomb repulsion between the charged ions, which at those high temperatures are completely stripped of their electrons, and to penetrate into the nuclear interaction region. It should be emphasized that the fact that the cross section for the fusion reactions diminishes for extremely high temperatures (higher than the maximum) provides an inherent safety against excessive heating of the fusion plasma.

Through the years there have been many debates on various schemes of magnetic confinement and many attempts have been made in the plasma laboratories to improve the performance of the machines: to increase the values of $n\tau_E$ to fulfil the Lawson criterion and at the same time obtain fusion temperatures has turned out to be an extremely difficult task.

7.3   The architecture of magnetic confinement

The architecture of the different magnetic confinement configurations has developed into an almost sculptural art. Configurations range from the first straight pinch currents and the first toroidal stellarator (Princeton), with the figure-of-eight variant, to today's tokamaks with D-shaped cross-section configurations like JET (Culham), new attempts with an elongated D cross-section (Lausanne) and new versions of stellarators like the Wendelstein VIIA (Garching), using highly structured magnetic coil windings. Superconducting coils are used for example in Tore Supra (Cadarache). They introduce a new approach to magnetic confinement technology and are being considered for the next generation of large fusion devices.

7.4   History of alternative concepts

Historically, the first experiments in the field of magnetic confinement were straight pinches, using high power current discharges to heat and to confine the plasma. The self-generated magnetic field from the current itself was responsible for the confinement. Such a system turned out to be unstable and needed an external longitudinal magnetic field combined with a conducting wall for stabilization.

In order to minimize particle losses caused by leaking along the magnetic field lines the next step was to bend the magnetic field lines to form a magnetic bottle or a torus. However, the curvature of the magnetic field lines introduced new problems. Strong externally produced toroidal magnetic fields were necessary to stabilize the plasma, for example in a tokamak. As a result the β value of the ratio between the thermal pressure and the magnetic pressure became rather low, a drawback for tokamak plasmas.

Another type of configuration, the reversed field pinch, is similar to a tokamak in that it has the shape of a torus carrying a toroidal current. It is different, however, in the sense that the toroidal magnetic field is much weaker than in a tokamak so that on average the toroidal and the poloidal magnetic fields are of the same order. In the reversed field pinch the direction of the magnetic field has a strong variation in the radial direction, so-called magnetic shear. Simultaneously, use is made of conducting walls to improve stability. The toroidal field usually maximizes in the centre and reverses sign near the wall. It is characteristic for reversed field pinch operation to first show an initial turbulent evolution of the plasma developing into a quasi-steady state in which the plasma experiences a slow temperature increase accompanied by a simultaneous decay of the density. The energy confinement time of the reversed field pinch is usually smaller than for tokamaks but the β values are considerably higher, of the order of 30–40%.

Another concept of magnetic confinement, the Spheromac, does not use any external coils at all to enclose the plasma current.

The toroidal and poloidal magnetic fields are of the same order of magnitude. The control of the plasma to establish an equilibrium is left free for the plasma itself, which results in some technical advantages. The spheromac concept may be considered similar to an astrophysical confinement system which self-organizes its plasma equilibrium with a minimum amount of energy.

7.5 Stellarators and tokamaks

The stellarator principle, which is still one of the main candidates for a future reactor, was originally developed in Princeton, USA, according to ideas of the astrophysicist Lyman Spitzer (see plate 13).

In a stellarator the poloidal as well as the toroidal magnetic fields are generated by external coils. This has the advantage of giving a certain flexibility in the control of magnetic properties. A stellarator has no current flowing in the plasma, which gives the important advantage that it can operate continuously. The absence of a current in the plasma also eliminates a free source of energy that could drive instabilities.

Important progress in recent years has made the stellarator concept a possible candidate for a future reactor. Two large machines are in early phases of operation: the Wendelstein VII–X in Garching, Germany, and the Large Helical Device (LHD) in Toki, Japan (see plate 13 and figure 7.1).

The tokamak concept has, however, provided the base for the most important developments in recent years and has become the greatest hope for realizing a fusion reactor (see plates 14 and 16). The important results in JET (Joint European Torus, Culham, UK) and TFTR (Thermonuclear Fusion Test Reactor, Princeton, USA) and many other tokamaks in the world have considerably advanced our level of understanding of the properties and behaviour of fusion plasmas and have increased technological experience to such an extent that hopes are high that fusion power plants producing electricity may be realized in the next century.

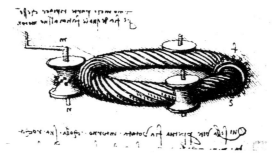

Figure 7.1. 'Perpetual screw'. Drawing by Leonardo da Vinci (Codex Madrid I).

The magnetic confinement principle has established itself as a useful method to substitute for gravitational fusion plasma confinement that the rest of the universe exploits for its power generation.

References

[7.1] Kadomtsev B B 1992 *Tokamak Plasma: a Complex Physical System* (Bristol: Institute of Physics Publishing)

[7.2] Wagner F 1997 Topics in toroidal confinement *Plasma Phys. Control. Fusion* **39** A23

Chapter 8

Microballoon explosions by lasers: inertial confinement

8.1 The principle of inertial confinement

The idea of using microballoon explosions to produce fusion energy relies on the principle of using the inertia of imploding targets to provide confinement. The targets are usually small spherical microballoons filled with deuterium–tritium (D–T) gas (1 mg cm^{-3}). The inner part of the small pellet contains the main fuel D–T region surrounded by an ablator. When energy is supplied from a driving laser the ablator heats up and expands, forcing the rest of the shell to contract inward to conserve momentum. Inertial confinement fusion uses pulsed high power lasers or ion beams as drivers. The pulses have to be so strong and short that the target does not have time to expand during the period of fusion production. The confinement time is so short, less than 10^{-10} s, that light does not propagate more than 1 cm over the course of the explosion. The particle density in the central region, where the fusion reactions occur (radius less than a tenth of a millimetre) is extremely high at 10^{25} cm^{-3} [8.1].

The fusion reactions begin in the central part of the pellet which contains a few per cent of the D–T fuel and propagates as a fusion burn-front in the rest of the fuel to complete the process. The details of the implosion are important for the efficient operation of the fusion burn. To avoid hydrodynamic instabilities, restrictions must be observed concerning the minimum pressure (100 megabars) and irradiance (10^{15} W cm^{-2}) during the pulse. Other effects like the symmetry of the implosion are also important for the ignition.

8.2   Instabilities

Hydromagnetic instabilities of the Rayleigh–Taylor and Kelvin–Helmholtz type for example may occur in the fuel when domains of different densities meet each other. In the nonlinear regime interaction between modes may lead to turbulence, which is studied intensively using numerical simulation methods.

Parametric instabilities may occur when laser radiation interacts with the natural modes of oscillation of the plasma. Nonlinear coupling with ion sound waves may generate stimulated Brillouin scattering and be accompanied by filamentation of the plasma, whereas interaction with electron plasma waves may result in stimulated Raman scattering. It may be interesting to mention that similar phenomena have been observed when high-power radio waves interact with the plasma in the ionosphere which surrounds the Earth; experiments on this have been carried out in the north polar region in connection with observations by means of the European Incoherent Scatter Facility (Kiruna,Tromsø).

8.3   The concepts of direct and indirect drive

There are two different paths which have been developed for coupling the energy from the driver to the target: one is known as direct drive, the other as indirect drive. For direct drive the laser beam or ion beam is incident directly on the fusion target pellet. Indirect drive is characterized by the laser or ion beam first being absorbed in a 'hohlraum', an enclosure of high atomic number surrounding the pellet. X-rays are then produced which act as a driver for the implosion.

Each type of driver has its advantages and its drawbacks. Direct drive transfers the energy directly to the pellet but is sensitive to the spatial quality of the radiating beams, whereas indirect drive is less sensitive in this respect and has the advantage that the implosion becomes hydrodynamically less unstable. Indirect drive is, on the contrary, more sensitive to laser-driven parametric instabilities, which may be generated in the plasma produced by the x-rays [8.2, 8.3]. Active research has resulted in detailed knowledge about the requirements for efficient operation of directly [8.4] and indirectly driven inertial confinement fusion.

8.4   Laser fusion

Large facilities like the Livermore experiments have used neodymium-glass lasers producing 1.05 μm radiation, which is commonly converted to 0.35 μm at the third harmonic to minimize instabilities from laser–plasma interaction. Experimental results from the Livermore Nova using indirect drive show excellent agreement with predictions from sophisticated numerical simulations of neutron yield, ion temperature and fuel density. Direct drive experiments have also shown considerable progress in many laboratories all over the world.

A fusion power plant based on inertial confinement consists of the following parts, namely a *driver* (laser or particle accelerator providing the fusion target with energy), a *target factory* (manufacturing the targets and filling them with D–T fuel), a *reactor* (the base for fusion energy generation by microexplosions) and a *generator* (producing electricity from thermal energy). The inertially confined plasma stays for only an extremely short time in a burning state (10^{-10} s) and is very small (10^{-2} cm) [8.5].

The repetition rate of pulses from the driver is a few pulses per second. The repetition pulse rate can be varied as can the yield of the target. This is advantageous since in development tests the experiments can also be carried out at moderate cost by using lower powers.

The targets should be prepared with a high surface finish to obtain high gain. The requirement is less for indirect drive. The targets are prepared in special drop towers which can produce hundreds of drops per second with nearly perfect spherical shells. Sophisticated techniques are being developed for filling the drops with D–T fuel, using, for example, diffusion filling in high-pressure chambers.

There are several attractive features which make inertial confinement fusion a worthwhile option to consider for a future energy producing plant. Development of megajoule lasers for experiments and tests of ignition and gain in reactor-size devices would be useful in this respect. The progress of advanced research in recent years has led to plans for constructing megajoule lasers for future inertial confinement experiments.

The technique of short-pulse laser radiation has advanced considerably in recent years. It has become routine to produce 10 TW pulses and to focus them to 10^{20} W cm^{-2}. As a result of the high power, a terawatt pulse in a plasma may produce self-channelling and excite ultrastrong (hundred gigavolt per metre) longitudinal electric fields and multimegagauss (100 MG) magnetic fields in the wake of the propagating pulse! Not only may the electronic component be excited, but also ion currents of the order of mega-amperes with ion energies of the order of 150 keV may be generated. Such energies are of interest in fusion, as the resonance peak in the cross section of D–T reactions happens to be not far away from this value. Self-generated 100 MG magnetic fields could pinch relativistic electrons causing laser filamentation and self-focussing.

New possibilities for simulating astrophysical systems in the laboratory are being opened up by the extraordinary power of high-intensity lasers. For the first time laboratory investigations of the magnetohydrodynamics of supernovas and of the radiative properties of stars and nebulae could be made under conditions which were hither to only to be found out in the cosmos.

Densities and temperatures corresponding to those of stellar interiors may be achieved in the laboratory with high-power lasers and used for studies of the opacity of dense matter. Relativistic astrophysical plasmas may, furthermore, be simulated with ultra-short-pulse lasers. Phase transition of ultradense hydrogen under high pressures, for example metallization or magnetization, may occur in the interiors of stars and planets, and may be basic mechanisms in the formation of supernovas. There are indeed many fascinating possibilities for discovery using high-power and ultra-short-pulse laser technologies. They open up a new field of research—a new link with the physics of the cosmos [8.6].

8.5   Ion beam drivers

Ion beam drivers use short beam pulses from either light-ion or heavy-ion accelerators to drive the fusion target, and a variety of pulsed power techniques have been developed for each type of acceleration. It turns out that heavy-ion accelerators offer particular advantages such as high-rate pulse production and reliability. Certain ions like xenon, caesium and bismuth accelerated to some GeV are well suited for absorption in fusion targets. For singly charged ions a current of about 100 kA is needed to obtain the required power of 10^{14}–10^{15} W. Experiments using either radiofrequency accelerators or induction accelerators are performed to obtain such high powers with beams of the required brightness.

The beam intensity needed to convert the accelerated beam energy to x-rays in indirect drive targets is about 10^{15} W cm^{-2}. The focal spot should be no larger than some millimetres, which requires a spread of less than 1% in the longitudinal momentum of the beam. More experiments are needed, possibly with megajoule drivers, to provide a base for future reactor development. With regard to the targets there are many features common to x-ray production in indirect drive experiments by lasers and by ion beams.

Heavy-ion drivers are attractive because of the high energy loss of the ions with distance of penetration into matter. To the lowest order this loss is proportional to Z^2, where Z denotes the ion charge state. According to the principle of indirect drive, the heavy ion beam is stopped in the converter, where its kinetic energy is transformed into x-ray radiation. The x-rays propagate through the 'hohlraum' and drive the fusion target isotopically.

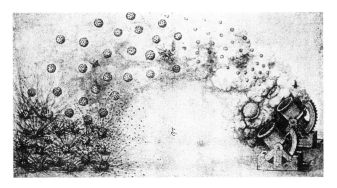

Figure 8.1. 'Explosive cannon balls ('cotombrots') ejected from cannons'. Drawing by Leonardo da Vinci (Codex Atlanticus 1502).

The interaction of the heavy ions and matter, which produces a hot plasma state of solid material density, is a complicated area. Recent measurements and simulations show deviations from the predictions of linear theory. Particle in cell (PIC) simulations made in 1997 indicate that nonlinear screening of the free electrons causes the observed weakening of the stopping power. A full understanding of related problems is important for target design, a crucial point in inertial confinement fusion.

The construction of efficient projectiles and their interaction with targets has been of interest for centuries. Leonardo da Vinci made a striking drawing of cannon balls which appears in figure 8.1. He gave them the strange name 'cotombrots'; they were half a foot wide and full of small projectiles, producing the most disastrous effects when thrown among enemies.

References

[8.1] Lindl J D, McCrory R L and Campbell E M 1992 Progress toward ignition and burn propagation in inertial confinement fusion *Phys. Today* September 32

[8.2] Kruer W L 1988 *The Physics of Laser Plasma Interactions* (Redwood City, CA: Addison-Wesley)

[8.3] Baldis H A and Labaure C 1997 Interplay between parametric instabilities in the context of inertial confinement fusion *Plasma Phys. Control. Fusion* **39** A51

[8.4] Nishimura H 1997 High-convergence uniform implosion of fusion pellets with the new GEKKO laser *Plasma Phys. Control. Fusion* **39** A401

[8.5] Hogan W J, Bangerter R and Kulcinski G L 1992 Energy from inertial fusion *Phys. Today* September 42

[8.6] Mourou G A, Barty C P J and Perry M D 1998 Ultrahigh-intensity lasers: physics of the extreme on a tabletop *Phys. Today* January p 22

[8.7] Lontano M, Mourou G, Pegoraro F and Sindoni E (ed) 1998 *Superstrong Fields in Plasmas: 1st Int. Conf. (August–September 1997, Varenna, Italy) (AIP Conf. Proc. 426)* (New York: American Institute of Physics)

Chapter 9

The fusion reactor

9.1 Reactor requirements

What can we say at present when it comes to the real goal of our efforts i.e. a fusion reactor or a fusion plant to produce electricity? What are the predictions and how can we formulate the requirements for the first reactor? There has been much consideration of these questions for the different schemes of magnetic confinement fusion and inertial confinement fusion. Since the most advanced plans have been developed for the tokamak reactor concept of magnetic fusion, let us take this as an example and summarize the conclusions.

Self-sustained fusion (ignition) requires the fusion triple product of density, energy confinement time and temperature to be greater than a certain value:

$$n\tau_E T > 5 \times 10^{15} \text{ (cm}^{-3} \text{ s keV).}$$

Separate requirements for the reactors in the fusion triple product emanate from conditions on the individual processes of fuelling a sufficiently dense plasma of deuterium and tritium

$$\text{density } n \sim (2\text{--}3) \times 10^{14} \text{ cm}^{-3}$$

heating to a sufficiently high temperature

$$\text{temperature } T \sim 20 \text{ keV } (200 \times 10^6 \, ^\circ\text{C})$$

and achieving sufficient thermal insulation

energy confinement time $\tau_E \sim 1–2$ s.

Even if in various experiments each of the conditions has been fulfilled separately the simultaneous fulfilment of all three necessary for ignition still lies in the future.

9.2   Reactor design

The first reactor is likely to be a high-power device, producing a few gigawatts of electric energy, operating with full ignition. It will be equipped with auxiliary heating and plasma control. It will also have an exhaust for helium ash and impurities. It may have a major radius R of about 8 m and a minor radius of 3 m. The toroidal magnetic field at the major radius should be of the order of 6 T and the plasma current 25–30 MA. Simulations by computer have been done for such a reactor and show that ignition can be obtained. Strict requirements are, however, necessary for the efficiency of the exhaust and impurity control systems.

An enormous number of data based on technology and physics predictions have been collected and prepared for the design of the ITER International Thermonuclear Reactor; this will be capable of producing tokamak physics data regarding confinement properties, disruptions, beta-limits, divertor performance and particle heating and confinement for reactor scale devices, and will be able to conduct nuclear testing at a neutron wall loading of about 1 MW m^{-2} [9.1, 9.2].

9.3   Heating and confinement

The next generation of large machines is being designed for auxiliary heating and current drive systems: Neutral Beam Injection (NBI), Ion Cyclotron Resonance Heating (ICRH), Electron Cyclotron Resonance Heating (ECRH) and Lower Hybrid (LH). Since none of these systems can fulfil all the operational requirements one attempts to increase the flexibility by using combinations of auxiliary systems in order to study different scenarios for optimizing performance.

The use of external heating has a tendency to lead to a degradation of plasma confinement. It has been found that such effects are extremely sensitive to the plasma conditions near the plasma external boundary. Detailed studies of related phenomena led to the discovery of a new mode, the so-called H-mode (high mode) of operation, in the ASDEX machine (Garching) which was later confirmed in other machines. Experiments with the H-mode, which was associated with steep gradients near the boundary, showed that heating could be obtained without degradation of the confinement. In fact, the H-mode could typically have twice the confinement time of that with L-mode (low-mode) operation.

The discovery of the H-mode may be seen as an example of how important improvements and new openings in science often come in steps. One reason for building larger and larger tokamaks is that the confinement time crudely depends on the tokamak minor radius a and the diffusion coefficient D as a relation which applies for energy as well as for particles and indicates that larger machines should provide better confinement. For practical and economic reasons one might, however, envisage other lines of development using stronger magnetic fields leading to higher densities n in more compact machines. One might even for such devices eliminate or reduce the external heating, using essentially ohmic heating produced by the plasma itself.

There are obviously a great number of scenarios it is possible to exploit, and the last word has by far not been said with regard to the optimum path for construction of a future reactor.

References

[9.1] Zweben S J *et al* 1997 Alpha particle physics in the tokamak fusion test reactor DT experiment *Plasma Phys. Control. Fusion* **39** A275

[9.2] Ono M *et al* 1997 Plasma transport control and self-sustaining fusion reactor *Plasma Phys. Control. Fusion* **39** A361

Chapter 10

Outlook into the future

Fusion will continue to be the energy source of the whole universe. The stars, among them our Sun, will continue to shine for billions of years. In the centre of the Sun the average time for a proton to undergo a fusion reaction with another proton to produce a deuterium is of the order of 10^{11} years, which makes the whole proton cycle extremely slow and the time scale for the consumption of the proton fuel immensely long. On a related time scale new stars will be born under the influence of gravitational forces, creating new systems of planets and in the process of formation possibly also new life.

It should not be forgotten that it is and will continue to be the Sun, by its own fusion processes, which continues to support the energy for life on our planet. Modern society may, however, affect the situation in our atmosphere by changing its composition in such a way that appreciable effects may be felt by life on Earth.

In our everyday life one encounters urgent questions about energy and one is even reminded that very serious energy problems may occur in the near future. The problems may seem more or less serious depending on where on Earth one is living, in one country or another, rich in resources or not, in a highly developed state or on developing continents, in highly populated areas or in areas which are scarcely populated.

However, when serious energy needs are faced, they will soon be felt all over the world, demanding common effort and requiring global solutions. It should be emphasized that in a world with rapidly increasing population and developing societies with increasing living standards, ecological problems will strongly influence the course of the future.

From extensive studies of the global ecology it has been reported that the greenhouse effect, which results in successive increases in atmospheric temperature, above what would normally be expected, could become a serious global problem. This effect is due to increasing quantities of CO_2 and other gases in the atmosphere which prevent escape of excess heat from the Earth. The greenhouse effect may be responsible for 'global warming' which could cause increases in sea level of several metres, due to melting of the polar ice-caps, causing population redistribution from low-lying areas.

There is indeed a real need for the development of an energy source which solves the large-scale energy problems of the world with due regard to possible ecological problems and damage. Fusion would provide a large-scale energy source for the production of electricity without the risks of explosion or other damaging effects and without negative ecological consequences. It does not give rise to any products that would contribute to an increase in the greenhouse effect. *The advantages of having fusion available as a power source on Earth would in fact be so immense that they would motivate all the necessary scientific, technological and economic support required for realization.* Great efforts will be needed to achieve the 'goal of fusion' in the time we have available before other options begin to run down. Science, technology and economics as the controlling parameters are all powerful tools enabling us to reach the goal.

The question of when a fusion reactor plant for the production of electricity could be expected to operate for the first time is often raised. The response could be: when do we need it? This depends on the balance between expected consumption and the available resources. The question of the necessary economic support to realize the construction of a fusion reactor is relevant, but not in fact decisive, since if it became urgent the necessary amount of support would almost certainly be provided. The development cost would necessarily depend on the risk one would be prepared to take with respect to successful operation of the reactor, i.e. to what extent the reactor would be optimized.

It is estimated that the next generation of large fusion experiments can be regarded as an extrapolation of previous experiments of the tokamak type, in particular of JET (Joint European Torus), which in 1991 was the first machine in the world to demonstrate nuclear fusion reactions in a plasma. JET has been considered a great technological success as well as one of international collaboration.

How large will an operating fusion plant be? Power production is planned to be about 1500 MW which is higher, but not excessively higher, than an ordinary nuclear reactor plant. It can well be incorporated into the power grids of, for example, the EU.

The technological and scientific success of fusion energy programmes has been achieved by international collaboration, in Europe as well as at the intercontinental level. One should not forget that sometimes technical innovations and technical efforts proceed faster than expected. So, for example, the first moon landing was accomplished earlier than projected. It was, however, even if very spectacular and impressive, a technical achievement less complex and less far-reaching for mankind than the realization of a fusion reactor would be. The reason for the complexity of a fusion experiment is the element of the plasma, the *enfant terrible* of the experiment, which makes the design of a fusion machine much more difficult than that of a particle accelerator for example.

It is, in fact, an inherent property of plasmas that they refuse to be governed by external forces. As an example, if ultra-high-power radiation in a laser fusion experiment is focussed on a pellet the created plasma immediately, due to strong nonlinear effects, tries to prevent the radiation from penetrating. One should, accordingly, not expect the plasma in nuclear fusion reactions to be particularly easy to handle, as is indeed the experience with fusion plasma.

However, when the game comes to an end the benefits will be enormous and Nature will help to provide a peaceful source of energy as it has done for eternity in the rest of the universe.

Conclusions—the great fusion plant

A tour of the universe has made it evident what a decisive role fusion energy plays on a large scale. Studies of the Sun, the solar wind, comets, the magnetosphere around our Earth as well as astrophysical and astrochemical observations of remote sources by means of optical instruments and radio telescopes have revealed many events and interrelations between phenomena which occur in the cosmos. We still do not know anything about the conditions of stellar objects belonging to planetary systems of other stars. Perhaps there are many more planets apart from our Earth where life exists and societies prosper from fusion energy produced by their 'mother stars' which could deliver heat by radiation to their planets as our Sun does to the Earth. Or maybe fusion energy plants have long since been developed on other planets by ingenious principles and skilful engineering.

As far as our own future is concerned it would not be surprising if most of the electrical power on Earth, let us say by the middle of the next century, were to be provided by fusion reactors. It should be emphasized that all alternative methods of generation of electricity on Earth, wind energy, wave energy from the sea, solar radiation converted by solar cells, etc, are all indirectly derived from the energy emitted by the Sun, i.e. they originate from solar fusion. Even the atmosphere, the rivers and the forests providing other energy alternatives for electric power are driven by heat and light from solar fusion. With a large-scale view of the cosmos as a whole one may simply conclude that not only are the Sun and all stars fusion reactors but that the entire universe is a great fusion plant.

Afterword

I spent my student years at Chalmers University of Technology in my home town of Göteborg in Sweden, where I also prepared my doctoral thesis on 'The scattering of electromagnetic waves by electron beams' (1959) under the direction of Professor Olof Rydbeck, founder of the Onsala Space Observatory. Among his numerous pioneering scientific contributions he discovered CH radical line radiation in galactic clouds at centimetre wavelengths. For many years he was a great stimulus to me in my scientific efforts.

My interest in fusion plasma physics was very much stimulated by my stay, in 1960–1, as a post-doctoral fellow in the Plasma Physics Laboratory in Princeton, USA, headed by Lyman Spitzer the famous astrophysicist. I met many of the leading plasma physicists in the US and I also wrote my first paper there on nonlinear waves in hot plasmas. That was the beginning of what came to occupy me for many years and to bring to me many happy episodes in my life.

Of the many personalities that I met in connection with fusion plasma activities the memory of Hans Otto Wüster (Director of the Joint European Torus, JET, from its very beginning) remains with me; he was a man of outstanding quality. In the early days of JET the USA was ahead of Europe in fusion research; this, however, did not scare Wüster who said that we could do it even better in Europe ourselves. Not very many years later experience proved that his prediction was correct. He was a genius at coordinating technical science and human relationships, a partnership which carried JET to the highest achievement in fusion activities. He made me and many others believe in Europe and in the advantage of coordinating the activities of European countries.

The contacts with scientists from different fields who I met in the Royal Swedish Academy of Sciences in Stockholm (as a member since 1974) have always been of particular interest to me. His Majesty King Carl XVI Gustaf of Sweden takes a particular interest in the activities of the Royal Academy, being also the Protector of the Academy. His presence at meetings and conferences in natural sciences, for example in ecology or plasma physics, has always been highly appreciated by the members. Through the years I have had several opportunities to participate in such conferences where the King was present.

In 1981 the European Incoherent Scatter Facility (EISCAT) based in Kiruna, Sweden, north of the Arctic Circle, and with collaborating stations in Tromsø, Norway, and Sodankylä, Finland, was inaugurated by King Carl Gustaf in the presence of Professor Bengt Hultqvist, the Director of the Swedish Space Center in Kiruna (formerly the Kiruna Geophysics Institute, KGI, of which the author was a Board Member for many years), and many scientific representatives. The King was also on the morning plane from Kiruna to Stockholm the next day. I had a connecting flight in Stockholm to Italy for the Varenna conference on plasma physics. The connection between the flights was very tight and the plane arrived in Stockholm late. I did not know how I was going to get on the plane to Italy in time. An unexpected source of help solved my problem. Did it come from God, from the King or from the airline company? I got a message, however, that I could be taken by the King's escort car directly on the runway from the plane from Kiruna to the Alitalia plane. Avoiding all formalities saved me time and I took off for Italy, where in Varenna at Lake Como the same day I had the opportunity to discuss problems on the plasma universe with Hannes Alfvén and furthermore to consider the planning of the 1982 International Conference on Plasma Physics to be held in Göteborg the next year. On that occasion King Carl Gustaf gave an inaugural address, as did Professor Kai Siegbahn from Uppsala University, President of the International Union of Pure and Applied Physics, and winner of the Nobel Prize for Physics in 1981. Hannes Alfvén gave an invited introductory talk, addressing the problems of the plasma universe. The conference was also attended by several top scientists from the Soviet Union, among them R Z Sagdeev, V L Ginzburg and A G Sitenko, visitors rather uncommon in those days, and by about 100 participants from the USA, among them M N Rosenbluth, J M Dawson and A Hazegawa together with more than 500 participants from 40 other countries. When a similar conference was held in Prague, Czech Republic, in the summer of 1998 the number of participants had more than doubled, which indicates the increased interest in the field.

In 1989 King Carl Gustaf attended the graduation ceremony when the new doctors from Chalmers University of Technology were inaugurated. As promoter I gave a lecture entitled 'Fusion: dream and reality?'. I remember the performance well since it was in May 1989, only a couple of weeks after the propaganda for 'cold fusion', a misleading announcement that nevertheless for some time seriously disturbed the reputation of fusion research in hot plasmas. The King got to know not only about the recent progress in magnetic confinement fusion but was also informed about the misleading concept of cold fusion at this early stage by the limerick I dedicated to the 'inventors':

Two chemists from Salt Lake City
Who wanted to be thought very witty
Said they had the solution
By means of cold fusion
But it didn't work out—what a pity!

(freely translated from the original Swedish).

And the King commented on it humorously at the banquet, when he talked about 'fusion' and 'not fusion'.

In 1974 the first Nobel prize related to astrophysics was awarded to the British scientists Martin Ryle and Antony Hewish for investigations on radioastronomy antenna systems with cosmological implications, and on pulsars, respectively. The same year the Swedish authors Eivind Johnson and Harry Martinson received the Nobel prize for literature for their novels and poetry. The Academy had given me the responsibility for introducing the work of the physics prizewinners to the audience in the concert hall in Stockholm. I remember that I asked Harry Martinson, who was sitting next to me behind the curtains just before the start of the ceremony, if he would agree to my quoting in my physics presentation his poetry on world clocks from his great opus *Aniara* in connection with pulsars. I was pleased to be told that he would be delighted. What a coincidence between physics and poetry!

In 1983 the physics of stars and the elements in the universe were topics for which Nobel prizes were awarded to S Chandrasekhar and W A Fowler. In his Nobel lecture at the Academy Professor Fowler, with his characteristic sense of humour, said 'Those of you who heard me in Göteborg the other day can go home. I will give the same lecture here'. I happened to come across Fowler on the campus after his lecture and told him 'I heard you give the same lecture more than 25 years ago at the Bohr Institute in Copenhagen, in 1957, when I was a student there'. Fowler responded with a big friendly smile 'With Niels Bohr sitting in the front row puffing his pipe!'. Fowler invited me to lunch together with Chandrasekhar. Chandrasekhar was going to come to Göteborg in the next few days, giving a lecture there and visiting our institute at Chalmers. From conversation with Chandrasekhar I learnt how as a young student he came to England and informed Sir Arthur Eddington, the leading British astronomer, about his own work on the critical solar mass. He said that Eddington had been nice and had accepted his ideas, which were in fact revolutionary.

Chandrasekhar came to Chalmers and spent a whole day there. His general lecture had the title 'Newton, Shakespeare and Bach'. I remember that I introduced him to the audience by telling the story that as a Professor of the University of Chicago he drove a long way every week from the observatory to give lectures to only two graduate students. However, one day he was greatly rewarded when his two students T D Lee and C N Yang, in 1957, shared the Nobel prize for physics. Now, their master got his award! Before the end of his visit to Chalmers Chandrasekhar was interviewed by the editor of the monthly student journal of the physics department. Chandrasekhar apparently enjoyed the interview. At the very end of it the student said that he knew that Chandrasekhar had spent four years writing his last book on black holes (he had by the way then already written six other books and about 600 scientific publications), and the question was if it was really worth it. Chandrasekhar hesitated a little then he said 'I can tell you that it gave me even more pleasure than this prize'. He was a real scientist!

Every year in December when the Nobel prizewinners come to Stockholm to receive their prizes they present their work in Nobel lectures. As a rule the lectures cover the highlights of their discoveries for which they have been awarded the prize. One year, in 1978, Academician Piotr Kapitza, at the age of 84 already then a legend among physicists, came from Moscow to receive his prize. That year, the physics prize was, by the way, shared between him and two other physicists from the US, namely Arno Penzias and Robert Wilson, who received their prizes for the discovery of the 3 K background radiation in the universe. Before coming to Stockholm Kapitza had already informed the perpetual Secretary of the Academy that he was not going to talk about his work on extremely low temperature physics and his discovery of superfluid liquid helium, the topic of his award. The reason was, as he said, that he had done that work 40 years ago, and that he had forgotten everything about it. So he was going to talk about his current interest in research which was *fusion*.

He started his lecture a little hesitantly in a rather timid voice, and I may not have been the only one in the audience to wonder how this would end up. But after a few minutes he apparently felt more and more at home, and as time went on he became even enthusiastic. His point of view was that thermonuclear fusion research had entered a state where there was not much hope of reaching a final solution. He said that a lot of effort in magnetic confinement was going on in the direction of tokamak experiments but, as he explained, tokamaks were too complicated and would not reach the goal. The other alternative with laser fusion was not very encouraging to him either, since the power of the available lasers was much too low to be attractive for fusion energy production. However, he himself had recently developed a new scheme for fusion plasma experiments and that was what he was going to talk about! And it was here that enthusiasm entered his talk. He emphasized that one should use the same process for confining the plasma and for heating the plasma. The idea he launched was to use microwaves for both purposes. His new experiments were already showing promising results!

Scholar on the prayer chair praying for strength and inspiration from Leonardo (by Carl Fredrik Reuterswärd).

A couple of years later I visited Moscow and the Kapitza Institute, where I had discussions with Professor Pytaevvsky, Kapitza's theoretician over many years of work. I was told that Professor Kapitza himself was not available. However, I would be welcome to visit the magnet laboratory. On my way there I happened by chance, as it seemed, to come across him in the corridor. He invited me to accompany him to his office, where, sitting at his desk, he very kindly informed me about his interests and his work. When he asked if there was something I would like to have from him in particular I asked if he could offer me a reprint of his Nobel lecture. Immediately, an assistant came from another room with a copy which Kapitza kindly signed. He then said 'This afternoon we shall do an experiment together'! He instructed his assistant to switch on the power for the microwaves. I got to see the bluish (as far as I remember!) faint plasma, only some centimetres wide, through a hole for diagnostic measurements. His experimental equipment was spread out on two floors of the laboratory, joined by a narrow spiral staircase. I shall never forget his enthusiasm, running up and down the stairs and trying not to stumble on the many wires that were on the floor! How different a scene from the computerized fusion machines operating today in the modern plasma laboratories. I remember Paul Henri Rebut, former head of the JET project once saying 'Plasma physics is not so nice anymore when we cannot *see* the plasma'. But remember: there will always be stars, stars, stars ... wherever we look throughout space.

Acknowledgments

It was a great challenge for me to write this book after my many years of interest in geocosmic plasma physics and fusion plasma physics. The outstanding recent developments in these fields provided a strong motivation. I was also particularly stimulated by discussions and questions raised when I gave talks in various places.

The beginning of the story was that Gerard Bonneaud invited me to give a talk on fusion to the elementary particle physics group at Ecole Polytechnique, Palaiseau, Paris. He encouraged me considerably during the early phases of the work. Several visits to the Istituto di Fisica del Plasma, Milan, and its director Giampietro Lampis became exceedingly fruitful for developing the contents of the book. I also greatly appreciated the comments from the audience at a talk in l'Academie Nationale des Sciences, Belles Lettres et Arts de Bordeaux.

Without my wife Julie the book would probably never have been finished. Her continual encouragement and help with the wordprocessing of the ever-changing manuscript have been invaluable.

Nils Robert Nilsson has been an unfailing supporter of my work. Our discussions on scientific, artistic as well as editorial questions have been more than appreciated and sometimes gave me courage to continue writing.

For the final form of the work generous and highly competent linguistic aid from my friend John Allsop has been particularly appreciated. He transformed my text into something that came as near as possible to his own Oxford English. He also made valuable constructive comments on the outline of the book.

Ros Herman for her part kindly helped me to improve the style of certain sections and Alexis Brandeker provided me with the beautiful picture he took of the Hale–Bopp comet in March 1997.

An original photograph of the recent painting *Cosmos* (1997) by Pierre-Marie Brisson was kindly put at my disposal by the artist.

Madame Evelyne d'Alblousse, Fondation Peter Stuyvesant, kindly offered me the use of a photograph of the extraordinary etching by André Masson *Les Oiseaux Sacrifiés* (1954, aquatint).

I would also like to thank the staff of Institute of Physics Publishing and in particular Michael Taylor for his great encouragement.

The Hubble Space Telescope and JET contributed with some outstanding photos from their activities, Garching Laboratories from the Astex Upgrade and stellarators, and Culham Laboratories from START. Material created with support to AURA/ST ScI from NASA contract NAS5-26555 is reproduced here with permission.

Joël Lafons, Laboratoire Photographie 'Speed Photo', Bordeaux, provided me with some delicate photographs of lightning strikes taken across the Bordeaux sky-line, one of them hitting the top of the cathedral spire!

Appendix 1

Here, we reproduce Hannes Alfvén's pioneering publication 'Existence of electromagnetic–hydrodynamic waves' in *Nature* (1942, issue 3805, 3 October, p 405). Here he introduces a new type of wave and describes the coupled motions of a conducting liquid and an electromagnetic field. The motions would become known as magnetohydrodynamics and the wave as the Alfvén wave.

Existence of Electromagnetic-Hydrodynamic Waves

IF a conducting liquid is placed in a constant magnetic field, every motion of the liquid gives rise to an E.M.F. which produces electric currents. Owing to the magnetic field, these currents give mechanical forces which change the state of motion of the liquid. Thus a kind of combined electromagnetic-hydrodynamic wave is produced which, so far as I know, has as yet attracted no attention.

The phenomenon may be described by the electrodynamic equations

$$\text{rot } H = \frac{4\pi}{c}\, i$$

$$\text{rot } E = -\frac{1}{c}\,\frac{dB}{dt}$$

$$B = \mu H$$

$$i = \sigma(E + \frac{v}{c} \times B);$$

together with the hydrodynamic equation

$$\partial \, \frac{dv}{dt} \;=\; \frac{1}{c} \, (i \, \times \, B) \;-\; \text{grad} \; p,$$

where σ is the electric conductivity, μ the permeability, ∂ the mass density of the liquid, i the electric current, v the velocity of the liquid, and p the pressure.

Consider the simple case when $\sigma = \infty$, $\mu = 1$ and the imposed constant magnetic field H_0 is homogeneous and parallel to the z-axis. In order to study a plane wave we assume that all variables depend upon the time t and z only. If the velocity v is parallel to the x-axis, the current i is parallel to the y-axis and produces a variable magnetic field H' in the x-direction. By elementary calculation we obtain

$$\frac{d^2 H'}{dz^2} \;=\; \frac{4\pi\partial}{H_0{}^2} \, \frac{d^2 H'}{dt^2},$$

which means a wave in the direction of the z-axis with the velocity

$$V \;=\; \frac{H_0}{\sqrt{4\pi\partial}}.$$

Waves of this sort may be of importance in solar physics. As the sun has a general magnetic field, and as solar matter is a good conductor, the conditions for the existence of electromagnetic-hydrodynamic waves are satisfied. If in a region of the sun we have $H_0 = 15$ gauss and $\partial = 0 \cdot 005$ gm. cm.$^{-3}$, the velocity of the waves amounts to

$$V \sim 60 \text{ cm. sec.}^{-1}.$$

This is about the velocity with which the sunspot zone moves towards the equator during the sunspot cycle. The above values of H_0 and ∂ refer to a distance of about 10^{10} cm. below the solar surface where the original cause of the sunspots may be found. Thus it is possible that the sunspots are associated with a magnetic and mechanical disturbance proceeding as an electromagnetic-hydrodynamic wave.

The matter is further discussed in a paper which will appear in *Arkiv för matematik, astronomi och fysik.*   H. Alfvén.

Kgl. Tekniska Högskolan,
    Stockholm.
        Aug. 24.

Appendix 2

Here, we reproduce Stig Lungquist's first simple demonstration of magnetohydrodynamic waves 'Experimental demonstration of magnetohydrodynamic waves' in *Nature* (1949, issue 4160, 23 July, p 145).

Experimental Demonstration of Magneto-hydrodynamic Waves

In an electrically conducting liquid placed in a magnetic field, there is a mutual interaction between electromagnetic and hydrodynamic forces. Alfvén[1] has shown that this interaction may cause a new type of waves, called magneto-hydrodynamic waves, which travel in the liquid in the direction of the outer magnetic field, carrying an induced magnetic field as well as a velocity field with them. It seems probable that they are of great importance in solar physics, especially for the theory of sunspots. They may be important also in other branches of cosmical physics. They have not hitherto been observed in the laboratory; but an experimental investigation is very desirable, especially as many problems connected with waves are very difficult to treat theoretically.

Due to the finite conductivity, the waves are damped very much under laboratory conditions, even in such a good conductor as mercury, and this makes it difficult to demonstrate the waves. The damping is decreased if the magnetic field is increased. Calculations show that it should be possible to observe the waves in mercury, subject to a magnetic field of the order of 10,000 gauss.

In the experiment, the mercury is enclosed in a cylindrical vessel with the axis parallel to a strong vertical magnetic field (see diagram). The diameter of the vessel is 15 cm. and the height is about 13 cm. The waves are excited mechanically by the rotational vibrations of a disk near the bottom of the cylinder. The tangential velocity, v, of the mercury at the bottom may thus be written $v = v_R \cdot \dfrac{r}{R}$, where v_R is the velocity at $r = R$, the radius of the cylinder. For the theoretical calculations this expression for v should be split up into a series of Bessel functions,

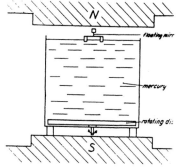

such that $v = \Sigma A_v J_1(k_v r)$, the wave-numbers being defined by $J_1(k_v R) = 0$. The waves corresponding to each term of this series are transported to the free surface of the mercury and are reflected there. Due to the finite conductivity, the damping and the wave velocity are different for different wave-numbers, causing a distortion of the wave.

A measuring device consisting of a small mirror floating on the surface is used for indicating the angular displacement. The mercury must be very clean, because the slightest trace of dirt at the surface brakes the motion of the mirror. Preliminary measurements of the damping and the phase-difference show a satisfactory agreement with the theoretical values. The frequencies used are 0·1–1 sec.$^{-1}$, and the outside magnetic field is about 10⁴ gauss. The amplitude damped to about 50 per cent and the phase-difference between top and bottom varies from 10° to 100. A detailed report will be given later.

Stig Lundquist

Department of Electronics,
Royal Institute of Technology,
Stockholm 26.
March 8.

[1] Alfvén, H., *Nature*, **150**, 405 (1942); *Ark. f. mat., astr. o. fys.* **29**, B, No. 2 (1942); **29**, A, No. 11 (1943); **29**, A, No. 12 (194 *Mon. Not. Roy. Ast. Soc.*, **105**, 3 and 382 (1945); **107**, 211 (194 *Ark. f. mat., astr. o. fysik*, **34**, A, No. 23 (1948). See also, Walén C., *Ark. f. mat., astr. o. fysik*, **30**, A, No. 15 (1944).

Appendix 3

On the following pages we reproduce Subrahmanyan Chandrasekhar and Enrico Fermi's brilliant and verifiable calculation of the strength of the magnetic field at our position in the spiral arm of our galaxy. They estimated it to be of the order of a millionth of a gauss in their classic paper 'Magnetic fields in spiral arms' in the *Astrophysical Journal* (1953, volume 118, p 116).

MAGNETIC FIELDS IN SPIRAL ARMS

S. Chandrasekhar and E. Fermi

University of Chicago
Received March 23, 1953

ABSTRACT

In this paper two independent methods are described for estimating the magnetic field in the spiral arm in which we are located. The first method is based on an interpretation of the dispersion (of the order of $10°$) in the observed planes of polarization of the light of the distant stars; it leads to an estimate of $H = 7.2 \times 10^{-6}$ gauss. The second method is based on the requirement of equilibrium of the spiral arm with respect to lateral expansion and contraction: it leads to an estimate of $H = 6 \times 10^{-6}$ gauss.

The hypothesis of the existence of a magnetic field in galactic space[1] has received some confirmation by Hiltner's[2] observation of the polarization of the light of the distant stars. It seems plausible that this polarization is due to a magnetic orientation of the interstellar dust particles;[3] for such an orientation would lead to different amounts of absorption of light polarized parallel and perpendicular to the magnetic field and, therefore, to a polarization of the light reaching us. On this interpretation of the interstellar polarization we should expect to observe no polarization in the general direction of the magnetic lines of force and a maximum polarization in a direction normal to the lines of force. And if we interpret from this point of view the maps[4] of the polarization effect as a function of the direction of observation, it appears that the direction of the galactic magnetic field is roughly parallel to the direction of the spiral arm in which we are located. In this paper we shall discuss some further consequences of this interpretation of interstellar polarization, in an attempt to arrive at an estimate of the strength of the interstellar magnetic field.

As we observe distant stars in a direction approximately perpendicular to the spiral arm, it appears that the direction of polarization is only approximately parallel to the arm. There are indeed quite appreciable and apparently irregular fluctuations in the direction of polarization of the distant stars.[4] This would indicate that the magnetic lines of force are not strictly straight and that they may be better described as "wavy" lines. The mean angular deviation of the plane of polarization from the direction of the spiral arm appears to be about $a = 0.2$ radians.[4] There must clearly be a relation between this angle, a, and the strength of the magnetic field, H. For, if the magnetic field were sufficiently strong, the lines of force would be quite straight and a would be very small; on the other hand, if the magnetic field were sufficiently weak, the lines of force would be dragged around in various directions by the turbulent motions of the gas masses in the spiral arm and a would be large. To obtain the general relation between a and H, we proceed as follows:

The velocity of the transverse magneto-hydrodynamic wave is given by

$$V = \frac{H}{\sqrt{(4\pi\rho)}} , \tag{1}$$

[1] E. Fermi, *Phys. Rev.*, **75**, 1169, 1949.

[2] W. A. Hiltner, *Ap. J.*, **109**, 471, 1949.

[3] Of the two theories which have been proposed (L. Spitzer and J. W. Tukey, *Ap. J.*, **114**, 187, 1951, and L. Davis and J. L. Greenstein, *Ap. J.*, **114**, 206, 1951), that by Davis and Greenstein appears to be in better accord with the facts.

[4] W. A. Hiltner, *Ap. J.*, **114**, 241, 1951.

where ρ is the density of the diffused matter. In computing the velocity, V, we should not include in ρ the average density due to the stars, since the stars may be presumed to move across the lines of force without appreciable interaction with them, whereas the diffused matter in the form of both gas and dust has a sufficiently high electrical conductivity to be effectively attached to the magnetic lines of force in such a way that only longitudinal relative displacements are possible.

According to equation (1), the transverse oscillations of a particular line of force can be described by an equation of the form

$$y = a \cos k\ (x - Vt), \tag{2}$$

where x is a longitudinal co-ordinate and y represents the lateral displacement. We take the derivatives of y with respect to x and t and obtain

$$y' = -ak \sin k\ (x - Vt)$$

and

$$\dot{y} = -akV \sin k\ (x - Vt). \tag{3}$$

From these equations it follows that

$$V^2\overline{y'^2} = \overline{\dot{y}^2}\ . \tag{4}$$

The lateral velocity of the lines of force must be equal to the lateral velocity of the turbulent gas. If v denotes the root-mean-square velocity of the turbulent motion, we should have

$$\overline{\dot{y}^2} = \tfrac{1}{3}\ v^2\ . \tag{5}$$

The factor $\tfrac{1}{3}$ arises from the fact that only one component of the velocity is effective in shifting the lines of force in the y-direction. The quantity y', on the other hand, represents the deviation of the line of force from a straight line projected on the plane of view. Hence,

$$\overline{y'^2} = a^2\ . \tag{6}$$

Now, combining equations (1), (4), (5), and (6), we obtain

$$H = (\tfrac{4}{3}\pi\rho)^{1/2}\ \frac{v}{a}\ . \tag{7}$$

In equation (7) we shall substitute the following numerical values, which appear to describe approximately the conditions prevailing in the spiral arm in which we are located:[5]

$$\rho = 2 \times 10^{-24}\ \text{gm/cm}^3, v = 5 \times 10^5\ \text{cm/sec, and } a = 0.2\ \text{radians.} \tag{8}$$

With these values equation (7) gives

$$H = 7.2 \times 10^{-6}\ \text{gauss.} \tag{9}$$

An alternative procedure for estimating the intensity of the magnetic field is based on the requirement of equilibrium of the spiral arm with respect to lateral expansion and contraction. As an order of magnitude, we may expect to obtain the condition for this equilibrium by equating the gravitational pressure in the spiral arm to the sum of the material pressure and the pressure due to the magnetic field. In computing the gravita-

[5] For ρ, the estimate of J. H. Oort (cf. *Ap. J.*, **116**, 233, 1952) from observations of the 21-cm line is used; while the value of v adopted is that of A. Blaauw, *B.A.N.*, **11**, 405, 1952.

tional pressure, we should allow for the gravitational force due to all the mass present, i.e., of the stars as well as of the diffused matter. We are interested, however, in computing the gravitational pressure exerted on the diffused matter only. Assuming for simplicity that the spiral arm is a cylinder of radius R with uniform density, one finds for the gravitational pressure:

$$p_{grav} = \pi G \rho \rho_t R^2 , \qquad (10)$$

where G denotes the constant of gravitation, ρ is the density of the diffused matter only, and ρ_t is the total mean density, including the contribution of the stars. The kinetic pressure of the turbulent gas is given by

$$p_{kin} = \tfrac{1}{3} \rho v^2 \qquad (11)$$

while the magnetic pressure is given by

$$p_{mag} = \frac{H^2}{8\pi}. \qquad (12)$$

And for the equilibrium we must have

$$p_{grav} = p_{kin} + p_{mag} . \qquad (13)$$

In computing p_{grav} we shall assume a radius of the spiral arm of 250 parsecs or $R = 7.7 \times 10^{20}$ cm. As before, we shall take $\rho = 2 \times 10^{-24}$ gm/cm³; and for ρ_t we shall assume[6] 6×10^{-24} gm/cm³. For these values of R, ρ, and ρ_t equation (9) gives $p_{grav} = 1.5 \times 10^{-12}$ dynes, while p_{kin} computed with the values already given is 0.2×10^{-12} dynes. We attribute the difference to the magnetic pressure. Hence

$$\frac{H^2}{8\pi} = 1.3 \times 10^{-12}, \qquad (14)$$

or

$$H = 6 \times 10^{-6} \text{ gauss.} \qquad (15)$$

The two independent methods of estimating H therefore agree in giving essentially the same value for the field strength. A field of about 7×10^{-6} gauss indicated by these estimates is ten times smaller than that which Davis and Greenstein[3] have estimated as necessary for producing an adequate orientation of the dust particles to account for the interstellar polarization. If the present estimate of 7×10^{-6} gauss is correct, one should conclude that the mechanism of orientation is somewhat more effective than has been assumed by Davis and Greenstein.

Since this paper was written, our attention has been drawn to the fact that the idea underlying the first of the two methods by which we estimate the magnetic field in the spiral arm is contained in an earlier paper by Leverett Davis, Jr. (*Phys. Rev.*, **81**, 890, 1951). We are sorry that we were not aware of this paper when we wrote ours. However, since with the better estimates of the astronomical parameters now available the value of H derived is a great deal different from Davis' value and since further the value we have derived is in accord with our second independent estimate, we have allowed the paper to stand in its original form.

[6] Cf. J. H. Oort, *Ap. J.*, **116**, 233, 1952.

Glossary

Mathematical notation for large and small numbers

Very large and very small numbers frequently appear in the text. The following concise notation is used.

10^n represents 1 followed by n zeros. In general, x^n means x multiplied by itself n times, e.g.

$$10^3 = 10 \times 10 \times 10 = 1\,000$$
$$10^6 = 1\,000\,000$$
$$3.2 \times 10^4 = 32\,000.$$

10^{-n} represents 1 *divided by* (1 followed by n zeros). In general, x^{-n} means 1 divided by x multiplied by itself n times, e.g.

$$10^{-3} = 1/10^3 = 1/1000 = 0.001$$
$$10^{-6} = 1/10^6 = 1/1\,000\,000 = 0.000\,001$$
$$3.2 \times 10^{-4} = 3.2/10^4 = 3.2/10\,000 = 0.00\,032.$$

Units

In scientific work an international system of units (*Systeme Internationale*—SI) is in general use.

The standard units and notation for length, time and mass are as follows.

Unit of length, 1 metre, denoted by 1 m
e.g. 3×10^3 m = 3000 metres

Unit of time, 1 second, denoted by 1 s e.g. 2×10^{-2} s = 2/100 second

Unit of mass, 1 kilogram, denoted by 1 kg e.g. 5.4×10^4 kg = 54 000 kilograms.

All other units of physical quantities can be expressed in terms of these basic units but, as illustrated below, are frequently given special names.

Other units which occur in the text are

Unit of speed, 1 metre per second, denoted by 1 m/s or 1 m s^{-1}

Unit of acceleration, 1 metre per second per second, denoted by 1 m/s^2 or 1 m s^{-2}

Unit of force, 1 kg m s^{-2} = 1 newton, denoted by 1 N

Unit of energy, 1 kg m^2 s^{-2} = 1 joule, denoted by 1 J

(In atomic, nuclear and particle physics, the electronvolt is generally used as the unit of energy. This is defined in the next section.)

Unit of electric charge, 1 coulomb, denoted by 1 C

Unit of electric potential difference, 1 volt, denoted by 1 V

Unit of electric current, 1 ampere (or amp), denoted by 1 A

Unit of electrical resistance, 1 ohm, denoted by 1 Ω

Unit of power, 1 watt, denoted by 1 W

Unit of rate of radioactive decay, 1 becquerel, denoted by 1 Bq

Unit of temperature, 1 kelvin, denoted by 1 K ($0\,^\circ$C = 273 K)

Fundamental physical constants

These are constants which play a key role in determining the scale and nature of the physical universe.

Constant	Symbol	Value
Speed of light	c	2.998×10^8 m s^{-1}
Planck's quantum constant	h	6.626×10^{-34} J s
$h/2\pi$	$\hbar$	1.055×10^{-34} J s
Gravitational constant	G	6.673×10^{-11} N m^2 kg^{-2}
Mass		
of electron	m_e	9.109×10^{-31} kg
of proton	m_p	1.673×10^{-27} kg
of neutron	m_n	1.675×10^{-27} kg
Proton charge	e	1.602×10^{-19} C

The following is the unit of energy generally used in atomic, nuclear and elementary particle physics:

electronvolt	eV	1.602×10^{-19} J
	keV = 10^3 eV	1.602×10^{-16} J
	MeV = 10^6 eV	1.602×10^{-13} J
	GeV = 10^9 eV	1.602×10^{-10} J

This glossary is adapted from Blin-Stoyle R J 1997 *Eureka! Physics of Particles, Matter and the Universe* (Bristol: Institute of Physics Publishing).

Short fusion–plasma dictionary

This 'dictionary' contains explanations of certain commonly used terms in the field of plasma and fusion physics. Indications are also given as to where those terms most frequently occur in the chapters of the book. The words are related to the cosmic plot of fusion and plasma terms shown in the Introduction.

Alpha particle An alpha particle is a helium nucleus consisting of two protons and two neutrons bound together by nuclear forces. Alpha particles are produced by fusion reactions between, for example, deuterium and tritium isotopes of hydrogen at the same time as high-energy neutrons are generated (chapters 2 and 7).

Aurora are spectacular coloured phenomena observed at high latitudes and are caused by the particles which enter the Earth's magnetic field from the solar wind (chapter 4).

Black hole Sufficiently massive stars may undergo gravitational collapse and become black holes which are even denser than neutron stars. One may imagine that nuclear matter has become completely crushed. In black holes the gravitational force is so large that no light can be emitted from them. The density is estimated to be 16 g cm^{-3} or about 100 times greater than that of a neutron star.

At the end of the 18th century Laplace, the French mathematician and physicist, proposed that heavenly bodies with the same density as the Earth and with a radius of 250 times the solar radius would have such a strong gravitational attraction that light could not escape from them and that the largest stars in the universe would remain invisible to us. Laplace, in his considerations, assumed, long before the concept of a 'photon' had been adopted, that 'light particles' could be attracted gravitationally according to Newton's law.

Chaos Chaos is a state of dynamic motion where chance plays a part but where certain regularities may remain (chapter 6).

Coherence The property of a set of vibrations or waves for which the mutual phases remain constant in time, a state of 'ordered motion'. The principle of lasers (light amplification by stimulated emission of radiation) is based on coherent light (chapter 8).

Comets Comets are objects in the solar system which often move in strongly elongated orbits. They consist of a nucleus of pieces of small frozen particles of gas and of stones, sand and dust. The nucleus is only some kilometres across. When comets approach the Sun evaporation occurs and forms a gaseous head to the comet and a tail of luminous gas several millions of kilometres long and often very spectacular. Plasma phenomena which occur in the ionized gas play an important role. Observation of radiation from comets, even recently observed x-rays, can give us information about the chemical structure of the dust, rock and ice which cannot have changed much since the birth of the solar system 4.5 billion years ago!

Creation The process of generating particles and radiation. It could increase the temperature of a plasma as expressed by a nonlinear term in the rate equation for the temperature in a plasma (T^p, where p is approximately 2) (chapter 6).

Density Characterizes the numbers of particles, for example electrons or ions, per unit volume in a plasma.

Deuterium Hydrogen isotope which consists of one proton and one neutron bound together by nuclear forces.

Diagnostics Measurements and evaluations of the plasma parameters (density temperature, current) as well as of phenomena occurring in the plasma (instabilities, fluxes of temperature and particle densities, radiation and particle emission, e.g. of fast neutrons). Methods of electron–cyclotron emission (ECE) and Thomson scattering of laser light are used to determine the electron temperature profiles in fusion devices.

Diffusion In a system of charged particles which collide with each other there will after many collisions, be a net migration of particles in space. The accompanying diffusion of energy and particles is a major problem in fusion research. The observed diffusion is generally much higher, about 100 times higher, than predicted by a classical collision model. The diffusion may be caused by collective oscillations associated with microinstabilities, for example temperature-gradient-driven drift-wave turbulence, leading to a particular temperature dependence of the thermal conductivity and particle diffusion coefficients, and $D \sim T^{3/2}$. The diffusion of the plasma determines the energy confinement time.

Earth Our planet, a very particular planet due to the presence of an atmosphere containing oxygen which we can breathe, and water in combination with temperatures which provide conditions for life. So far it is the only planet known to us which has these advantages!

Electric field Electric field originates from charged particles and affects the motion of other charged particles. An electric field always accompanies time-varying magnetic fields, for example those responsible for the toroidal current of interest in magnetospheric plasmas (chapter 4).

Electron Elementary particle with negative charge and mass $1/1836$ of the proton mass. The electrons occupying different energy levels in atoms are associated with different quantum numbers of spin and angular momentum, determining the frequency of radiation from transitions in the atomic structures. The free electrons in a metal are responsible for its electrical conductivity.

Energy Capacity of a system to produce work. For any system there exists an equivalence between mass and energy shown by the Einstein equation $E = mc^2$ where E is the energy, m the mass and c the velocity of light (3×10^{10} cm s^{-1}). The equation tells us that one gram of mass is equivalent to 25 million kilowatt hours. The liberation of energy by the transmutation of protons and neutrons to helium is accompanied by a mass defect which multiplied by c^2 gives the energy.

Equilibrium Balance of a system, which could be stable or unstable according to the consequence of a perturbation. A fusion plasma could exhibit several types of instability but might also stay sufficiently long in an equilibrium state for fusion reactions to produce useful energy (chapters 6 and 7).

Eruption Phenomenon where energy is released as a result of available free energy. Volcanic activities are perhaps the most spectacular example of eruptions at close distance. Eruptions associated with Sun spots generate various types of prominence. These are enormously energetic and spectacular phenomena. The spots occupy local domains on the Sun as large as the continents on Earth. They carry complex interweaved structures of magnetic fields and currents. The prominences could extend above the solar surface to heights up to one-quarter of the solar diameter (chapter 4).

Evolution Development in time and space of a system in general, for example the state of meteorological conditions, or the human population on Earth. The evolution of a hot fusion plasma to the burning plasma state in a fusion reactor under controlled conditions is a topic of considerable interest in fusion research. So is the evolution of the universe for cosmologists, or the evolution of laser communication by optical fibres, or of the Internet for consumers in general (chapters 4 and 6).

Explosion A state where a certain quantity, e.g. temperature grows nonlinearly in space and time to reach unlimited values in a definite time. This is what defines an explosive instability, a self-creating nonlinear effect. It is believed that prominences on the Sun are affected or even caused by such effects, which are also driving sources in a thermonuclear bomb explosion. For the theoretical description of a future fusion reactor the source terms will be balanced by terms describing diffusion processes and losses to result in an equilibrium operative state (chapter 6).

Flame A white-hot stream of luminous gas produced by combustion. Rocket flames are weakly ionized gas with electron densities of about 10^{12} per cubic centimetre and temperatures of near 1000 °C. If they are sufficiently large they can be classified as plasmas (Debye length λ_D less than $(T/n)^{1/2}$, the cross section of the flame). A rocket flame can cause black-out of a microwave signal of 3 cm wavelength and disrupt signal communications (chapter 4).

Future It is said to be difficult to make predictions, in particular about the future! The future is, in fact, the open domain in the direction of which everything evolves and where all is as yet unknown. It is the area for speculation for unexpected happenings and unknown trends, which change all deterministic expectations. It is a highly nonlinear domain!

Future cannot be predicted
it has to be invented

Dennis Gabor (awarded the Nobel prize for physics in 1971 for holography which he invented in 1948).

Fusion Nuclear reactions between light atomic nuclei such as isotopes of hydrogen from which heavier elements are produced with generation of large amounts of energy. Fusion is the source of energy in the universe and the origin of all light from stars and galaxies. The construction of fusion reactors is the ultimate goal of present-day fusion research (chapter 2).

Galaxies Large-scale structures of the universe containing on average 100 billion stars each and extending about 100 000 light-years, often in spectacular forms of flat spirals (chapter 3).

Instability Instability occurs in a plasma driven by some energy-releasing source which could be inhomogeneity in the temperature and/or density of the plasma or in the magnetic field as well as drift energy of a charged particle beam in the plasma. Different types of interacting wave can give rise to new waves for which the plasma provides conditions for growth and instability; one talks of nonlinear instabilities which can lead to turbulence where a large number of waves participate (chapter 6).

Ions Ions are the charged particles that remain when the atoms have lost or gained electrons, e.g. by radiation or collisions. They respond to low-frequency electromagnetic oscillations and to ion cyclotron wave magnetic fields (chapter 1).

Jet A plasma formation that may occur under different circumstances, from jet burners for the preparation of metal surfaces to elongated galactic jets which could have extensions over distances of many thousands of light-years (chapter 4, and plates 5 and 6).

JET Joint European Torus (Culham, UK). The large fusion plasma experiment based on the tokamak principle (chapter 7).

Laser Light amplification by stimulated emission of radiation has provided a new tool to be used in almost all branches of science and technology. Laser fusion is being investigated as one option for future energy production. Optical communication by fibre techniques is being revolutionized thanks to the advantages of using coherent light from lasers (chapter 8).

Life The origin of life and the biology of interstellar matter may be connected with suitable conditions created in mixtures of gas and dust. Radioastronomy attempts to analyse the presence of heavier molecules and their formation in the cosmos. The search for the presence of life on planets other than the Earth is a great challenge for the future.

Light The carrier of information in the universe which from ancient times has provided us with all the information we have about astrophysics and astrochemical phenomena and the rules governing the cosmos (chapter 3).

Magnetic field Magnetic fields are very important for confinement of fusion plasmas as well as for cosmic plasma phenomena in general (solar, galactic, magnetospheric, pulsars, comets, quasar physics). Magnetohydromagnetic plasma phenomena (MHD) are essential in laboratory fusion experiments as well as in cosmic plasmas (chapters 4 and 7).

Matter In the conventional sense the states of matter are solid, liquid and gaseous, these states being acquired at different levels of increasing temperature. For temperatures sufficiently high to cause ionization the plasma state is reached, the so-called fourth state of matter (chapter 1).

Nebula A luminous cloud of gas and dust occurring in interstellar space. The clouds, which most often have a dimension of some tens of light-years, often exhibit polarized magnetic fields generated by currents in the nebula.

Neutron Elementary particle carrying no net charge, which together with protons form the building blocks of atoms.

Neutron star Assumed to be formed by the collapse of a star and to be the final state of a supernova explosion. A neutron star may be as heavy as the Sun but only 10–15 cm in diameter. The density is enormous, about 10^{14} g cm^{-3}. A teaspoon of neutron star matter would weigh 100 million tons. The central temperature is exceedingly high, about 10 million degrees. Pulsars are believed to be rotating neutron stars, emitting repetitive signals of radiation.

Nonlinearity A phenomenon related to a high concentration of energy (particle or radiation density) in space and time. A wave of high amplitude could change the medium locally to such an extent that the wave propagates differently, for example with different speed and shape, from how it would have done in the absence of the self-induced conditions of propagation. The presence of other waves occurring simultaneously in the medium could change the conditions for a particular wave and they could also provide possibilities for the particular wave to create new waves due to mutual interactions. The advent of the laser in the early 1960s provided an excellent opportunity to create and also to study nonlinear phenomena in plasmas as well as in solids. Nonlinear effects play an important role not only in physics but in most modern fields of science as well as in the dynamics of our society (chapter 6).

Origin A point or region in space–time where a source of events is to be found. It might be possible to localize the origin of a fire but not the origin of the universe. One may say that it is everywhere (chapter 3).

Oscillation Vibrational motion of a charge particle, for example an electron, in a periodic electromagnetic field, a radiofrequency wave or an optical frequency field from a laser (chapter 6).

Photon A light-particle or quantum of radiation of energy $E = h\nu$, where ν is the frequency and h Planck's constant.

Plasma A gas where the atoms are ionized, i.e. where the electrons are separated from the ions by the influence of, for example, radiation or particle collisions. A plasma is electrically conductive, i.e. currents may occur in the plasma as a result of electric fields. Plasma is the most abundant state of matter in the universe. The word *plasma* is of Greek origin and means a state of matter with plastic properties (chapter 1). Plasma currents may generate magnetic fields, and magnetic fields may also influence the currents and control their motions (chapter 4).

Population Number of constituents in a state which is often in time-dependent evolution. A quadratic source dependence could generate 'explosive' tendencies (chapter 6).

Power Electrical energy generated per unit of time, for example in units of watts = joule/second. Future fusion reactors are expected to operate in the range of 3000 megawatts or three billion watts, which could be compared with the power of ordinary lamps of 60 watts, home electric radiators of 1000 watts or the 4×10^{23} kilowatts continuously emitted by the Sun in the form of light as electromagnetic waves (chapters 3 and 4).

Prominence Enormous luminous gas clouds ejected from the Sun demonstrating several types of plasma phenomenon such as instabilities, filamentation and blob formation, influence and generation of magnetic fields etc (chapter 4).

Propagation The process of continued change of position in time of a certain wave or packet of waves or pulses in a medium, for example a plasma. The speed of propagation is defined by the density and temperature of the plasma and by the presence of a magnetic field (chapter 6).

Proton Elementary particle with a mass 1836 times that of an electron and of positive charge.

Pulsars Sources of rapidly repeated pulses of radiation of frequencies from radio to visible and even x-ray or gamma-ray waves. The generation comes from neutron stars, objects of very high density which rotate with high velocity sometimes several hundred times per second, and have strong magnetic fields, sometimes thousands of billions of gauss. A pulsar exists in the centre of the Crab nebula generating pulsar radiation in the plasma from a supernova remnant (chapter 4).

Quasars Quasi-stellar radio sources at enormous cosmic distances from us, possibly 10 billion or more light-years away (chapter 4).

Radiation All charged particles which change their velocity, its absolute value in rectilinear motion (special relativity) or in direction (general relativity), produce electromagnetic radiation. Relativistic electrons which have a circular motion radiate electromagnetic radiation (synchrotron radiation) in their instantaneous direction of motion, which provides the radiation from all radio stars.

Synchrotron radiation also occurs in fusion plasmas when the fast electrons move in the magnetic field. The term radiation is often used to denote the corpuscular motion of particles (chapter 6).

Reaction Process of mutual direct influence between different elements, for example particles or molecules, to produce new elements. Nuclear fusion in stars is an example of a reaction which produces energy and alpha particles from protons. The hydrogen isotopes deuterium and tritium can react with one another to produce alpha particles and energy in a sufficiently hot gas mixture of these isotopes; this is the reaction process which would seem to be the most efficient for generating fusion energy in the future (chapters 2, 3, 7, 8 and 9).

Sky Synonymous with the cosmos or universe but usually imagined as the blue or starry half-sphere which hangs over our heads, reminding us of the mythological or magical views of ancient times.

Solar wind The permanent continuous flow of particles out of the Sun, predominantly electrons and protons, which reaches the Earth after about four days. It represents a link of plasma between the Sun and the Earth and has an important influence on the magnetospheric structure of the Earth as it feeds the magnetosphere with charged particles (chapter 4).

Source Origin of available free energy of any kind. As regards plasma and fusion physics it could be a laser source for laser–plasma interaction, resulting in the formation of small, dense, high-temperature domains by inertial confinement, or it could be a particle injection source for neutral beam heating of a magnetized fusion plasma (chapter 6).

Space Infinitely extended domain of which the centre is unknown.

Space–time Concept in the general theory of relativity based on four dimensions, i.e. the ordinary three spatial dimensions and in addition the time dimension.

Stellarator Toroidal confinement system where the toroidal drift of particles is compensated by external helical magnetic fields. In a stellarator there is no induced toroidal current flowing in the plasma. Unlike a tokamak, a stellarator can therefore operate continuously. The absence of plasma current also limits the possibilities for undesired instabilities.

Stars Luminous points in the sky which can be seen at night. They produce and emit energy and are born by contraction of interstellar matter. In the process of collapse the temperature rises to 10 million degrees, sufficient for the hydrogen nuclei to produce fusion energy and helium (alpha particles) (chapter 3).

State Physical system, for example solid, liquid, gas or plasma, or energy level of a particle or atom, for example ground state, where the electrons populate the lower levels of the atoms, or excited state, where the electrons populate higher energy levels defined by quantum numbers, or ionized (plasma) state where the atoms have been split into free electrons and ions.

Sun Our nearest star where fusion reactions are continually liberating large amounts of energy from which the Earth captures one part in 2 billion on average. It seems that the Sun's energy production has remained about the same for the last 3 billion years. The Sun is the most important element in our solar system since it supplies the energy necessary for life (chapter 3).

Sun spots Dark spots occurring on the Sun's surface often in groups. They are centres of magnetic activity with characteristic values of about 3000 gauss and are cooler than the rest of the surface. Noted by Galileo in 1610, Sun spots have still not been fully investigated as far as their electromagnetic properties are concerned (chapter 4).

Supernovas Sources of extreme brightness which occur when heavy stars undergo gigantic explosions (chapter 4).

Temperature Characterizes the heat of matter, for example of a plasma, where the temperature T is a measure of the disordered motions of the charged particles expressed in terms of their thermal velocity v_t, i.e. $kT = mv_t^2$, where m is the particle mass and k the Boltzmann constant.

Time A concept that helps us to determine the order of events.

Tokamak A toroidal chamber with magnetic coils. The tokamak magnetic confinement system consists of a toroidal field combined with a poloidal field produced by the current flowing in the plasma. The plasma current is produced by a large transformer, which of necessity has to operate intermittently. Coils around the central limb of the transformer core form the primary winding and the torus of the plasma itself is the secondary winding.

Tritium Hydrogen isotope which consists of one proton and two neutrons bound together by nuclear forces.

Turbulence Disordered motion among elements in, for example, a fluid where filaments and vortices become mixed instead of conserving their identity. Turbulence in a plasma may be regarded as a state composed of a large number of waves of different frequencies and wavelengths (chapter 6).

Universe The totality of space where all stars, galaxies, interstellar matter and magnetic fields are distributed. The universe is the gigantic laboratory where all the astrophysical and cosmological phenomena occur, and which may be observed from Earth or from remote space observatories. The possibility of realizing fusion energy on Earth increases as we obtain more information about fusion–plasma phenomena in the universe (chapter 3).

Wave Systematic propagating undulatory motion characterized by a wavelength determined by the distance between two successive equal phases of the repetitive motion, for example two minima or two maxima. Waves could be of acoustic, optical or radio types for example. In plasmas a large number of waves of different types, for example longitudinal plasma waves, transverse electromagnetic waves, magnetohydrodynamic waves, cyclotron waves etc, can exist and can be used for heating or plasma diagnostics (chapter 6).

Subject index

Character index